Felix Geiger

FAST-RESPONSE MEASUREMENTS OF ORGANIC TRACE SPECIES IN THE EARTH'S ATMOSPHERE

Fast-response measurements of organic trace species in the Earth's atmosphere

by
Felix Geiger

KIT Scientific Publishing

Dissertation, Karlsruher Institut für Technologie (KIT)
Fakultät für Physik, 2014

Tag der mündlichen Prüfung: 31. Oktober 2014
Referenten: Prof. Dr. Johannes Orphal, Prof. Dr. Thomas Leisner

Impressum

KIT Scientific Publishing

Karlsruher Institut für Technologie (KIT)
KIT Scientific Publishing
Straße am Forum 2
D-76131 Karlsruhe

KIT Scientific Publishing is a registered trademark of Karlsruhe Institute of Technology. Reprint using the book cover is not allowed.

www.ksp.kit.edu

Print on Demand 2015

ISBN 978-3-7315-0356-9
DOI 10.5445/KSP/1000046297

Fast-response measurements of organic trace species in the Earth's atmosphere

Zur Erlangung des akademischen Grades eines
DOKTORS DER NATURWISSENSCHAFTEN
(Dr. rer. nat.)

von der Fakultät für Physik
des Karlsruher Instituts für Technologie (KIT)
- Großforschungsbereich -

genehmigte

DISSERTATION

von

Dipl.-Phys. Felix Benedikt Geiger

aus Starnberg, Bayern

Tag der mündlichen Prüfung:	31.10.2014
Referent:	Prof. Dr. Johannes Orphal
Korreferent:	Prof. Dr. Thomas Leisner
Dekan:	Prof. Dr. Dieter Zeppenfeld

Abstract

Volatile organic compounds (VOCs) are trace organic species that originate from biogenic sources (like e.g. the metabolism of plants or putrefaction), all up to anthropogenic sources (like e.g. incomplete combustion or as volatile by-products from miscellaneous traffic-related and industrial processes). Benzene, acetone, methanol or isoprene are well-known species of this group. Although typically appearing only in smallest traces, VOCs affect air quality and climate processing in multiple ways. They play some key roles as reactants within atmospheric reaction channels, like e.g. the formation of ozone or emergence of organic aerosol. Some are additionally classified as hazardous air pollutants. Thus their investigation is needed to understand climatic processes and to assess their effects on environment and human health.

This work reports on the development and use of a fast-response high-sensitivity in situ instrument for measurements of VOCs in ambient air. The Ultra-Light-Weight Proton-Transfer-Reaction Mass Spectrometer (ULW-PTR-MS) bases on the ionization of organic compounds in a reaction chamber using hydronium ions and the subsequent detection of product ions by a quadrupole mass spectrometer. The setup is particularly characterized by its compact size and little weight of $55\ kg$ (compared to twice the weight of commercial instruments). The specifically developed control electronics base on a modular measurement and control system and allows fast adaptation to mission-specific requirements. The instrument is intended to be deployed on research and passenger aircraft and will be engaged during the mission Oxidation Mechanisms Observation (OMO) aboard the German research aircraft HALO in 2015. The mission will investigate the free-radical chemistry at higher altitudes during the summer monsoon in Asia.
The ULW-PTR-MS was first-time used for ground measurements during the air quality study Uintah Basin Winter Ozone Study 2012 (UBWOS 2012). The primary task of this mission was the examination of enhanced formation of surface ozone during wintertime in the surroundings of intense fossil fuel production in a mountain basin in the Western United States. Comprehensive characterization and calibration measurements in comparison to other high-sensitivity VOC instruments were performed that attested the high accuracy of the ULW-PTR-MS. Subsequent measurements aboard a laboratory van allowed the direct investigation of emission fingerprints of VOCs originating from oil and natural gas well pads. At a recently hydraulically re-fractured gas well, that was investigated in more detail, very high mixing ratios of aromatic hydrocarbons in the range of several $ppmV$ (parts per million by volume, 10^{-6}) were found. These values outnumber typical urban

compositions by several orders of magnitude. Differences in the emission fingerprints of aromatics were determined downwind of the condensate storage tanks and the open flow-back pond showing, that the tanks emit a lighter composition of aromatics (i.e. higher percentage of compounds with lighter molecular mass) and the pond a heavier composition. The deduced chemical fingerprints of the diverse emitters allow conclusions on their contributions to the air composition in the region.
The results presented in this work are some of the first measurements using fast-response VOC instruments to investigate individual emitters in this chemically unusual environment. In this way they filled in the blank between stationary measurements of background signals and mobile on site characterizations of point source emissions. Within the UBWOS 2012 measurements it was further showed that Proton-Transfer-Reaction Mass Spectrometry (PTR-MS) is suited for mobile measurements aboard a driving laboratory.

Zusammenfassung

Flüchtige organische Verbindungen (VOCs) sind organische Spurenstoffe, die sowohl aus biogenen Quellen (z.B. Stoffwechsel von Pflanzen, Fäulnisprozessen), wie auch aus anthropogenen Quellen (z.B. Produkte unvollständiger Verbrennung, flüchtige Nebenprodukte aus Industrie und Verkehr) hervorgehen. Benzol, Azeton, Methanol oder Isopren sind bekannte Vertreter dieser Stoffklasse. Obwohl sie typischerweise nur in kleinsten Spuren vorkommen, beeinflussen VOCs sowohl Luftqualität wie auch Klimaentwicklung auf vielfache Weise. Als Reaktionspartner spielen VOCs eine Schlüsselrolle innerhalb verschiedener atmosphärischer Reaktionszyklen, wie z.B. der Ozonbildung oder der Entstehung von organischem Aerosol. Einige sind zudem als gefährliche Luftschadstoffe klassifiziert. Die Vermessung von VOCs ist daher notwendig, um sowohl klimatische Prozesse zu verstehen, wie auch Effekte auf Umwelt und Gesundheit abzuschätzen.

Die vorliegende Arbeit berichtet über die Entwicklung und den Einsatz eines in Echtzeit arbeitenden hochempfindlichen in situ Messgerätes für VOCs. Das ultraleichte Protonen-Transfer-Reaktion Massenspektrometer (ULW-PTR-MS) basiert auf der Ionisation organischer Verbindungen mit Hydroniumionen in einer Reaktionskammer und der anschließenden Detektion der Produkt-Ionen in einem Quadrupol-Massenspektrometer. Das Gerät zeichnet sich insbesondere durch kompakte Größe und geringes Gewicht von $55\ kg$ aus (verglichen mit dem doppelten Gewicht kommerzieller Geräte). Die eigens entwickelte Steuerelektronik basiert auf einem modularen Mess- und Kontrollsystem und ermöglicht die Anpassung an missionsspezifische Anforderungen. Das Instrument ist zum Einsatz auf Forschungs- und Passagierflugzeugen vorgesehen und soll 2015 im Rahmen der Mission Oxidation Mechanisms Observation (OMO) auf dem deutschen Forschungsflugzeug HALO eingesetzt werden. Hierbei soll die Chemie freier Radikale in großen Höhen zur Zeit des Sommermonsuns in Asien untersucht werden.
Das ULW-PTR-MS wurde während der Luftqualitätsstudie Uintah Basin Winter Ozone Study 2012 (UBWOS 2012) erstmals für Bodenmessungen verwendet. Das primäre Ziel dieser Mission war die Untersuchung verstärkter Bildung von Ozon am Boden zur Winterzeit im Umfeld intensiver Förderung fossiler Brennstoffe in einem Talkessel im Westen der Vereinigten Staaten. In diesem Rahmen konnten umfangreiche Charakterisierungs- und Kalibriermessungen im Vergleich mit anderen hochempfindlichen VOC Messgeräten durchgeführt werden, die eine hohe Genauigkeit von ULW-PTR-MS dokumentierten. Die anschließenden Messungen an Bord eines mobilen Labors ermöglichten die Untersuchung von VOC Emissionen direkt an Öl- und Gasförderanlagen. Eine kurz zuvor hydraulisch gefrackte

Gasquelle wurde dabei genauer untersucht. Sehr hohe Konzentrationen von aromatischen Kohlenwasserstoffen im Bereich von einigen $ppmV$ (Teile einer Million, 10^{-6}) wurden gemessen. Diese Werte übertreffen typische städtische Konzentrationen um mehrere Größenordnungen. Windabwärts der Lagertanks und des Flowbackspeichers wurden unterschiedliche VOC Emissionscharakteristiken festgestellt. Hierbei konnte insbesondere gezeigt werden, dass die Emissionen aus den Tanks einen höheren Anteil von Verbindungen leichterer molekularer Masse aufwiesen, während der offene Flowbackspeicher Mischungen schwererer VOCs emittierte. Die ermittelten chemischen Fingerabdrücke der verschiedenen Quellen lassen Rückschlüsse auf deren Beitrag zur Luftzusammensetzung in der Region zu.

Die in dieser Arbeit vorgestellten Ergebnisse stellen die ersten Echtzeit-Messungen von VOCs an einzelnen Emissionsquellen in dieser chemisch ungewöhnlichen Umgebung dar. Damit schließen sie die Lücke zwischen stationären Hintergrundmessungen und vor-Ort Charakterisierungen von Emissionen einzelner Punktquellen. Im Rahmen der UBWOS 2012 Kampagne konnte ferner gezeigt werden, dass Protonen-Transfer-Reaktion Massenspektrometrie (PTR-MS) geeignet ist für mobile Messungen in einem fahrenden Labor.

Contents

List of Figures

List of Tables

Part I

Introduction: Atmospheric Trace Organic Compounds Measured By Proton-Transfer-Reaction Mass Spectrometry (PTR-MS)

Chapter 1

Volatile Organic Compounds (VOCs)

1.1 Organic trace gases in atmospheric chemistry

With its extremely complex network of chemical and physical processes, the Earth's atmosphere is mainly driven by solar radiation, interaction with other compartments (biosphere, land, ocean, etc.), occasionally by volcanic eruptions and increasingly by mankind. With today's knowledge about the structure of the atmosphere, its derived classification into different compartments and spheres according properties like temperature, chemistry or aerodynamics (i.e. Planetary Boundary Layer (PBL), free troposphere, tropopause, stratosphere, etc.) and the ability to measure atmospheric components and dynamics up to high altitudes, many processes can be explained on scales of local to global dimensions. The atmospheric composition as pool of reactants comprises the main constituents nitrogen ($\sim 78.1\,\%$), oxygen ($\sim 20.9\,\%$) and several noble gases like Argon ($\sim 0.9\,\%$) and others ($< 0.002\,\%$), that are supplemented with a variety of trace gases ($\sum \ll 1\,\%$). Solely these less abundant trace gases control climate, weather and air quality as they play crucial roles in the Earth's radiative balance and chemical processing. Besides natural variations also anthropogenic interferences affect and stimulate the atmosphere and with that the global atmospheric budgets of their components. With a spectrum ranging from below one second up to thousands of years, the chemical lifetimes of involved compounds (e.g. seconds for OH and hundreds of years for chlorofluorocarbons - CFCs) and with that the spatial scales of their impacts (e.g. meters for OH and tens of thousands of kilometers for CFCs) vary strongly. The various processes have been comprehensively described in standard references, like e.g. Brasseur & Solomon [1986]; Seinfeld & Pandis [2006].

Carbon as central component - organic chemistry in the atmosphere The existence of carbon in various forms is a basic prerequisite for the habitability of the Earth and a fundamental constituent of life in general. The group of volatile organic compounds (VOCs) generally includes all organic trace compounds in the vapor phase, but excludes the most abundant atmospheric (organic) gases methane (CH_4), carbon monoxide (CO) and carbon dioxide (CO_2) for historical reasons. As the term denotes, VOCs are organic molecules that are especially characterized by high vapor pressures and low boiling points, respectively. From the variety of possible combinations of gaseous carbonic molecules, the most abundant atmospheric VOC classes are alkanes (only single bonds), alkenes (with double bonds), cycloalkanes (containing rings), aromatics (containing aromatic rings) and oxygenated (containing oxygen) hydrocarbons [Atkinson, 2000; Atkinson & Arey, 2003a]. In addition, it should be mentioned that besides being environmental pollutants by their nature anyway, some VOCs are even classified as air toxics that can have severe effects to people's health [EPA, 2012].

Among the diverse constituents being involved in the various atmospheric processes, especially these organic trace compounds profoundly affect the atmosphere despite their low atmospheric concentrations, like e.g. acetone (CH_3COCH_3) with a mixing ratio of $\sim 500\,ppt$ [Seinfeld & Pandis, 2006] in the free Northern mid-latitudinal upper troposphere and an averaged lifetime of ~ 250 days [Arnold *et al.*, 2004].

Since being substantially involved in processes like the formation of O_3 (discussed in section 1.2.1, p. 6) and secondary organic aerosol (SOA) (discussed in section 1.2.3, p. 9) in polluted air, VOCs also govern various tropospheric chemical reactions as catalyst, play a role in the global carbon cycle [Heimann, 2010] and are generally an important factor concerning air quality issues. With this multifaceted occurrence, VOCs are essential elements playing some key roles in atmospheric chemistry.

Some examples of atmospheric VOCs can be found in table T 2.2 in section 2.2.2 (p. 16). More general compilations are given in the relevant literature, e.g. in Atkinson [2007]; Derwent [1995]; Seinfeld & Pandis [2006]; Warneke *et al.* [2007].

Atmospheric VOC sources Diverse sources emit VOCs into the atmosphere. With origins ranging from natural emissions during biogenic processes (i.e. metabolic processes in plants, putrefaction or decomposition) all-up to anthropogenic emissions coming from miscellaneous traffic-related and industrial processes (i.e. incomplete combustion or as volatile by-products, organic solvents), their wide-ranging impact is obvious. In total the biogenic emissions are thought to be globally at least ten times larger than all anthropogenic contributions [Williams, 2004]. The most important biogenic VOC sources are isoprene emitting plants, followed by oceans, the ground and all kinds of living matter, like people, plants, animals, fungi and microorganisms. On the other hand the dominating anthropogenic contributions to organic emissions comprise especially the exploitation and production of fossil fuels, followed by biomass burning and industry. A rough estimate of the global yearly VOC emissions of both biogenic and anthropogenic origins results to $\sim 1300\,Tg(\mathrm{C})$[1]$\cdot yr^{-1}$, whereby isoprene by itself already contributes with $\sim 500\,Tg(\mathrm{C})\cdot yr^{-1}$ [Goldstein & Galbally, 2007].

[1] $Tg(\mathrm{C})$: $10^{12}\,g$ Carbon

Atmospheric VOC sinks/removal The removal of VOCs from the atmosphere usually takes place via reactions that predominantly comprise processes with the hydroxyl radical (OH). The final products of this gas-phase oxidation are then CO_2 and H_2O. To a lesser extent also photolysis and chemical oxidations with O_3 especially for alkenes occur, that are supplemented by reactions with nitrate radicals (NO_3) during nighttime [Atkinson & Arey, 2003a; de Gouw, 2005; Lewis *et al.*, 2005; Warneke, 2004]. Ketones (e.g. acetone - CH_3COCH_3) and aldehydes (e.g. formaldehyde - HCHO) are primarily degraded by photolysis. Likewise, dry and wet depositions and uptake by atmospheric aerosols (see section 1.2.3, p. 9) reduce VOC concentrations [de Gouw & Warneke, 2007].

1.2 VOC impacts on Air Quality and Atmospheric Processing

VOCs are present in virtually all atmospheric environments. In the following three crucial VOC reaction channels, that are connected to this work, are briefly described and demonstrate some key roles regarding air pollution and influences onto the climate via indirect effects. The first complex of themes covers tropospheric O_3 chemistry in connection with VOC involvement and is related to performed air quality field measurements in Utah/USA (see part III, p. 95). The second example describes the scientific framework of future measurements in the context of the oxidative capacity of the atmosphere, which is subject to upcoming airborne deployments of the system on passenger (see section 5.1, p. 83) and research (see section 5.2.2, p. 88) aircraft. The third example emphasizes the necessity for PTR-MS investigations in the context of the formation of secondary organic aerosol (SOA), which is one of the big uncertainties in future climate predictions.

1.2.1 Topic 1: VOCs in the tropospheric ozone cycle

Ozone (O_3) is omnipresent in both stratosphere and troposphere. Stratospheric O_3, which is formed at high altitudes ($\sim 15 - 50\,km$) from the dissociation of molecular oxygen in the UV by sunlight (Chapman cycle), shields the earth from incoming solar UV radiation ($\lambda < 360nm$) [Matsumi & Kawasaki, 2003]. Tropospheric ozone on the other hand is an air pollutant, that can have severe respiratory health effects [Fiore *et al.*, 2012]. As there is no direct emitter of O_3, the tropospheric ozone is either due to stratospheric entrainment of ozone-rich air, or due to in situ photochemical formation. Thereby the only significant tropospheric formation route begins with the photolysis of nitrogen dioxide (NO_2):

$$NO_2 \xrightarrow{h\nu} NO + O(^3P) \qquad (R\,1.1)$$

with $h\nu$ representing the energy of the solar radiation (h: Planck's constant[2]; ν: Frequency). The subsequent termolecular reaction of electronically excited $O(^3P)$ (from reaction R 1.1) with molecular oxygen and O_2 or N_2 molecules (M) forms ozone:

$$O(^3P) + O_2 + M \longrightarrow O_3 + M \qquad (R\,1.2)$$

O_3 (from reaction R 1.2) can then react with NO (from reaction R 1.1):

$$O_3 + NO \longrightarrow NO_2 + O_2 \qquad (R\,1.3)$$

regenerating NO_2, but at the same time constituting an ozone forming and destroying cycle that has a net ozone formation of zero.

However, the regeneration process of nitrogen dioxide can differ from the standard reaction R 1.3 by converting NO to NO_2 via radical catalyzed oxidations of

[2] $h = 6.62606876(52) \times 10^{-34}\,J \cdot s$ [Hollas, 2009]

(i) organic hydrocarbon species (and VOCs in general) (represented by R) or (ii) carbon monoxide (CO). Corresponding reactions with the hydroxyl radical ($OH^\bullet$) typically are:

$$(\mathrm{i}_1) \qquad \mathrm{RH} + \mathrm{OH}^\bullet \longrightarrow \mathrm{R}^\bullet + \mathrm{H_2O} \qquad \text{(R 1.4)}$$

$$(\mathrm{i}_2) \qquad \mathrm{R}^\bullet + \mathrm{O_2} + \mathrm{M} \longrightarrow \boxed{\mathrm{RO_2^\bullet}} + \mathrm{M} \qquad \text{(R 1.5)}$$

$$(\mathrm{ii}_1) \qquad \mathrm{CO} + \mathrm{OH}^\bullet \longrightarrow \mathrm{CO_2} + \mathrm{H}^\bullet \qquad \text{(R 1.6)}$$

$$(\mathrm{ii}_2) \qquad \mathrm{H}^\bullet + \mathrm{O_2} + \mathrm{M} \longrightarrow \boxed{\mathrm{HO_2^\bullet}} + \mathrm{M} \qquad \text{(R 1.7)}$$

The formed organic peroxy ($RO_2^\bullet$) radicals (reaction R 1.5) and $HO_2^\bullet$ radicals (reaction R 1.7) both convert $NO^\bullet$ to NO_2:

$$(\mathrm{i}_3) \qquad \mathrm{RO_2^\bullet} + \mathrm{NO}^\bullet \longrightarrow \mathrm{RO}^\bullet + \mathrm{NO_2} \qquad \text{(R 1.8)}$$

$$(\mathrm{i}_4) \qquad \mathrm{RO}^\bullet + \mathrm{O_2} \longrightarrow [\mathrm{carbonyl}] + \mathrm{HO_2^\bullet} \qquad \text{(R 1.9)}$$

$$(\mathrm{ii}_3) \qquad \mathrm{HO_2^\bullet} + \mathrm{NO}^\bullet \longrightarrow \mathrm{OH}^\bullet + \mathrm{NO_2} \qquad \text{(R 1.10)}$$

The carbonyl group in R 1.9 comprises e.g. ketones (RCOR) or aldehydes (RCHO).

The oxidation of organic hydrocarbons via reactions R 1.4 and R 1.5 results in the regeneration of a nitrogen dioxide according reaction R 1.8. The formed $RO^\bullet$ can further react producing another NO_2 molecule in combination with a hydroxyl radical, which is then available for future oxidation cycles. Reaction R 1.10 represents the final conversion of the $HO_2^\bullet$ radical. With both reactions R 1.8 and R 1.10, that complement the nitrogen dioxide regeneration according reaction R 1.3 in the sense of O_3 preservation, the net total ozone concentration increases. Reviewed laboratory studies of the underlying reaction pathways can be found in Orlando & Tyndall [2012].

Acting as fuel for the photochemical production of O_3, VOCs contribute to tropospheric/surface ozone formation (see also part III, p. 95). Thus, the entire process can be summarized as:

$$\text{VOCs + sunlight + NO}_x \rightsquigarrow \text{Ozone (+ other products)}$$

Ozone research activities Ozone research in troposphere and stratosphere has been extensively performed over years [Atkinson & Arey, 2003b; Crutzen, 1974; Jacob, 2000; Lelieveld *et al.*, 2008; Singh *et al.*, 1995] and represents one of the important topics in the recent IPCC reports. Although the atmospheric concentrations of ozone-destructing CFCs have been expected to phase out after the statement of the Montreal protocol in 1987, this field again attracts attention since so far unknown CFCs have been detected during CARIBIC aircraft measurements reported by Laube *et al.* [2014]. Especially over the Northern hemisphere partially even increasing concentrations have been observed, that cannot be neglected (e.g. CFC-216ca: 2013 concentration $20.2 \pm 0.3\,ppq$, increasing at current rate of $0.2\,ppq\cdot year^{-1}$ [Kloss *et al.*, 2014]).

Likewise, ozone as air quality issue has been identified to exist in the context of fossil fuel exploration and production during wintertime conditions [Carter & Seinfeld, 2012; EPA, 2012; Pinto, 2009; Rappenglück *et al.*, 2014; Schnell *et al.*, 2009]. Current research activities within the scope of the Uintah Basin Winter Ozone Studies (UBWOS) determined a significant contribution of radical amplification reactions to the formation of tropospheric radicals in such unique chemical environment (in addition to the above-described reaction channels; see also 1.2.2, p. 8), which facilitated an enhanced formation of surface ozone. The photolysis of carbonyl compounds (oxygenated VOCs with a C=O moiety) has been found to be the dominant source of oxidant (further discussed in section 7.1.1, p. 113). [Edwards *et al.*, 2013, 2014]

1.2.2 Topic 2: VOCs as precursors of HO_x

The self-cleansing of the atmosphere via oxidation of various trace gases (including organic species with CO and CH_4 is governed by HO_x radicals (HO_x=OH+HO_2) and to a far lesser extent by the reaction of some species with O_3 or by reaction with NO_3 during nighttime. That is, the mixing ratio of HO_x determines the chemical lifetimes of virtually all VOCs [Prinn, 2003].

HO_x formation The primary tropospheric HO_x source is photolysis of ozone and subsequent reaction with water vapor [Folkins & Chatfield, 2000]:

$$O_3 \xrightarrow{h\nu} O(^1D) + O_2 \quad \text{(R 1.11)}$$

$$O(^1D) + H_2O \longrightarrow 2\,OH \quad \text{(R 1.12)}$$

At higher altitudes the decreasing water concentrations lead to a decreasing contribution of this process R 1.12. With that other OH producing channels can then provide significant contributions. In this context, the known suitable organic secondary OH precursors, including acetone (CH_3COCH_3), methyl hydroperoxide (CH_3OOH) and formaldehyde (CH_2O), then increase the oxidizing power [Folkins & Chatfield, 2000]. Regulating factors are again ozone and nitrogen oxides (NO_x=NO+NO_2)

[Neumaier *et al.*, 2014] (see section 1.2.1, p. 6). The subsequent degradation of the organic hydrocarbon again follows reaction R 1.4. Relevant oxidation paths and reaction schematics are further discussed in Atkinson & Arey [2003a]; Monks [2005].

Research activities A study presented by Lelieveld *et al.* [2008] reports, that enhanced biogenic VOC emissions over the pristine Amazon forest did not deplete OH in the way it was expected. The particular atmospheric oxidative capacity was not reduced, which indicates that other photochemical processes in this unpolluted air (i.e. low NO_x region) recycle the OH efficiently well through further reactions of organic peroxy radicals. A recently published study by Fuchs *et al.* [2013] reports about investigations performed in the atmosphere simulation chamber SAPHIR[3] at the Research Centre Jülich showing that OH was additionally recycled without NO involvement in isoprene-rich environments, such as forests. The hydroxyl radicals, which have been detected in significantly higher concentrations than expected from model calculations, were strongly enhanced due to this additional recycling of radicals. Unimolecular reactions of isoprene peroxy radicals thus contributed significantly to the oxidation efficiency in the atmosphere. It was further assumed, that this class of recycling reactions plays a role in the oxidation of other unsaturated VOCs (like e.g. methacrolein) [Crounse *et al.*, 2012] and hence can be key to the further understanding of high OH concentrations in environments with high VOC concentrations.

The scheduled aircraft mission OMO (Oxidation Mechanism Observations) will further investigate the HO_x chemistry within the context of the oxidative capacity of the atmosphere (further described in section 5.2.2, p. 88).

1.2.3 Topic 3: Secondary Organic Aerosol

Improving the knowledge about the fate of organic carbon in the atmosphere is one of the major objectives for future VOC research: ending up in oxidation products and/or as aerosol particles.

Organic compounds as part of the atmospheric particulate mass The largest fractional amount of tropospheric particulate mass (PM) ($20 - 90\,\%$ of $PM_{submicron}$) consists of organic material [Jimenez *et al.*, 2009], that is either directly emitted into the atmosphere, i.e. is of primary origin (e.g. through direct release of particle mass from sources such as motor vehicles and forest fires [Robinson *et al.*, 2007]), or is formed from gas-phase precursors containing carbon, i.e. is of secondary origin (e.g. through photooxidation of biogenic precursors [Claeys *et al.*, 2004]).

Especially the secondary organic aerosol (SOA) is still insufficiently quantified to estimate its affection to the global radiation budget. The broad variety of natural and

3 `http://saphir.fz-juelich.de/`

human-made sources of VOCs, by which organic particulates with higher molecular masses can be formed, and the spatial distributions of their emissions with temporal variabilities and chemical transformations make an assessment more difficult [de Gouw & Jimenez, 2009].

Sources of SOA The basic SOA formation comprise the oxidation of VOCs via radicals of OH and NO_3, O_3 or Cl, forming less volatile products. The subsequently partitioned aerosol particles can then react to higher mass products. Dominant pathways of SOA production from atmospheric VOCs are broken down in Ziemann & Atkinson [2012].

Impacts of SOA Aerosols influence the development of the Earth's climate through direct (scattering and absorption of solar radiation) and indirect (serving as cloud-condensation nuclei) impacts. According to the Intergovernmental Panel on Climate Change (IPCC), aerosols (besides clouds) are one of the biggest uncertainties in future climate predictions [Riipinen *et al.*, 2012].

Besides their direct and indirect climatic impacts aerosols have strong effects on the quality of air and can cause atmospheric environmental pollution and severe health effects to people [Pöschl, 2005].

Research activities Especially the formation of secondary organic aerosol originating from both natural and man-made sources has recently attracted growing attention. In connection with air quality in mega-cities (discussed e.g. in Borbon *et al.* [2013]; Ryerson *et al.* [2013]), environmental impacts after damaging events, like e.g. the explosion of the oil platform Deepwater Horizon [Bahreini *et al.*, 2012; de Gouw *et al.*, 2011; Li *et al.*, 2013; Middlebrook *et al.*, 2012] or general assessments of organic contributions to the atmospheric environmental development, like e.g. in Fiore *et al.* [2012]; Robinson *et al.* [2007], trace organic gases and their secondary products are currently under intensified investigation.

As PTR-MS allows the characterization of SOA precursors in real-time, the use of this in situ technique contributes to these examinations.

Chapter 2

Principles of Proton-Transfer-Reaction Mass Spectrometry (PTR-MS)

Proton-Transfer-Reaction Mass Spectrometry (PTR-MS) is an on-line mass spectrometric technique for measuring organic trace gases. Since these gases take part in innumerable processes (not only atmospheric), the PTR-MS technology is used in widespread fields of application, ranging from the use in medical and biological fields, chemical analysis, food and flavor sciences, questions in security and crime scenarios, monitoring of general and specific industrial processes up to atmospheric concerns towards air quality issues.

The following introduction to PTR-MS is oriented on standard literature including some of the pioneering and method establishing works by Hansel *et al.* [1995]; Lindinger & Jordan [1998]; Lindinger *et al.* [1993, 1998] and especially de Gouw & Warneke [2007], and can be studied therein in greater detail. Likewise, reviews of the method can be found in Blake *et al.* [2009]; Hewitt *et al.* [2003]. The introduction to quadrupole mass spectrometry techniques and applications follows Dawson [1976]; Douglas [2009].

2.1 Methodology and Technology

Chemical Ionization Mass Spectrometry (CIMS) Ions of different mass can be molecularly weighed in the gas phase using mass spectrometry. Chemical ionization (CI), which is one of the applied ionization techniques, bases on ion/molecule reactions in a flow drift tube. Due to the corresponding lower energy processes compared to electron ionization, CI yields less fragmentation and with that mostly intact molecular species in form of protonated molecules. The idea of chemical ionization (CI) has been introduced by Munson & Field [1966], and the swarm technique of the flow drift tube has been described in a series of publications by McFarland [1973a,b,c].

Proton-Transfer-Reaction Mass Spectrometry (PTR-MS) PTR-MS, as branch of CIMS, is a relatively novel analytical detection technology for on-line monitoring of VOCs. This in situ technique allows identification and quantification of compounds at low concentrations (as low as parts per trillion by volume ($pptV$): 10^{-12}) in real time. Basis of PTR-MS applications is the soft chemical ionization, which is premised on the transfer of protons from a donor molecule to the molecule to be analyzed. The standard reagent ion hydronium (H_3O^+) can be formed from water vapor in a hollow cathode discharge. The subsequent analysis of the product ions has been extended from using a quadrupole mass spectrometer (QMS) (see section 2.3, p. 26) to other mass spectrometric methods. These are time-of-flight[4] (ToF) [Graus *et al.*, 2010; Jordan *et al.*, 2009b] and ion trapping[5] mass filtering [March, 1997; Prazeller *et al.*, 2003; Warneke *et al.*, 2005]. This lineup of filtering techniques is once more expanded by the most recent instrumental development presented by Sulzer *et al.* [2014], which is the combination of a quadrupole as an ion guide (instead of transfer lenses) and ToF for the detection.

By now also ions of nitric oxide (NO^+), molecular oxygen (O_2^+), krypton (Kr^+) and xenon (Xe^+) are used as precursor ions [Jordan *et al.*, 2009a], but the ionization method for these additional reagent ions is charge-exchange (CTR) rather than proton-transfer (PTR) ionization. However, the combination of these selectable reagent ions enhances the number of detectable substances [Edtbauer *et al.*, 2014; Sulzer *et al.*, 2012].

Instrumental configuration used in this work As the PTR system presented in this work is intended to perform atmospheric measurements of VOCs mainly aboard aircraft, the instrumental setup has been chosen such that especially prerequisites concerning airborne applications are met (see section 5, p. 83). Therefore a PTR system that uses hydronium precursor ions and a quadrupole mass filter in front of the detector has been chosen due to the following considerations:

- **Precursor ions - H_3O^+:** Distilled water as base material for the generation of H_3O^+ ions is an aeronautically unobjectionable substance, that is available practically everywhere. It allows an easy safety approval in contrast to otherwise needed gas cylinders. The same applies for toxic (e.g. ammonium - NH_3) or radioactive (e.g. polonium) materials, that could also be used for ionization demands. Moreover, H_3O^+ allows the analysis of a broad majority of important VOCs in atmospheric processes. Nonetheless, the combined use of several reagent ions within a single instrument, as mentioned above, allows the measurement of different varieties of compounds even with isomeric[6] separation. Therefore an application of this setup has to be revised regularly according to the specific scientific questioning.

[4] Time-of-flight mass spectrometry: mass-to-charge ratio evaluation of ions according their flight times in an electric field

[5] Ion Trap mass spectrometry: controlled storage of ions in electro-magnetic ion traps (e.g. Penning or Pauli) and subsequent evaluation according the ions' mass-to-charge ratios

[6] isomer: same sum formula

- **Mass filter - Quadrupole:** Quadrupole filters typically allow the design of smaller instruments compared to ToF-systems and are more robust and with low weight. These key aspects for airborne applications oftentimes decide about actual deployments (depending on the individual platform). However, the meanwhile advancements in mass spectrometry and product design referring to these progresses open further options for VOC measurements and enlarge the optional lineups of instruments. The broader potential for high-sensitivity measurements with even separation of isobaric compounds[7] in time-of-flight or ion-trapping systems dramatically widen the informative level of measurements, but do not automatically guarantee further insights. Since the operational areas covered with the instrument described in this work comprise mainly measurements of chemical processes in the atmosphere (middle to upper troposphere and lower stratosphere) (described in section 5, p. 83), where only a limited variety of volatile organic compounds is crucially involved, a quadrupole mass spectrometer can sufficiently well determine the observed processes. Measurements with other objectives on the other hand definitely benefit strongly from these enhanced technical features, but oftentimes have to be purchased at the expenses of size and weight (at least for ToF-systems).

Schematic view of the PTR system Figure F 2.1 (p. 14) presents the schematic overview of the PTR setup. A dashed line schematically illustrates the ion trajectories through the instrument from the IS to the detector. The main components of the instrument are:

- **Ion reactor:** Ion source (IS) to produce H_3O^+ precursor ions; Ion source drift region (ISDR) as intermediate reaction chamber; Drift tube (DT) reaction chamber, where the PTR between H_3O^+ and the injected ambient air VOCs take place (depicted with H_3O^+/RH^+)
- **Intermediate vacuum chambers (IVC):** Intermediate chambers with several pumping stages to enable the transfer and injection of ions into the high-vacuum chamber
- **QMS**: High-vacuum spectrometer with quadrupole mass filter and secondary electron multiplier (SEM) to measure the reagent and product ions

The chemical/physical processes of this PTR system are described in the next section 2.2 (p. 15), whereas specific descriptions of the setup of the newly constructed instrument are given in the second part of this work (see part II, p. 35).

---

7 isobaric components: exactly same nominal mass, but different sum formula

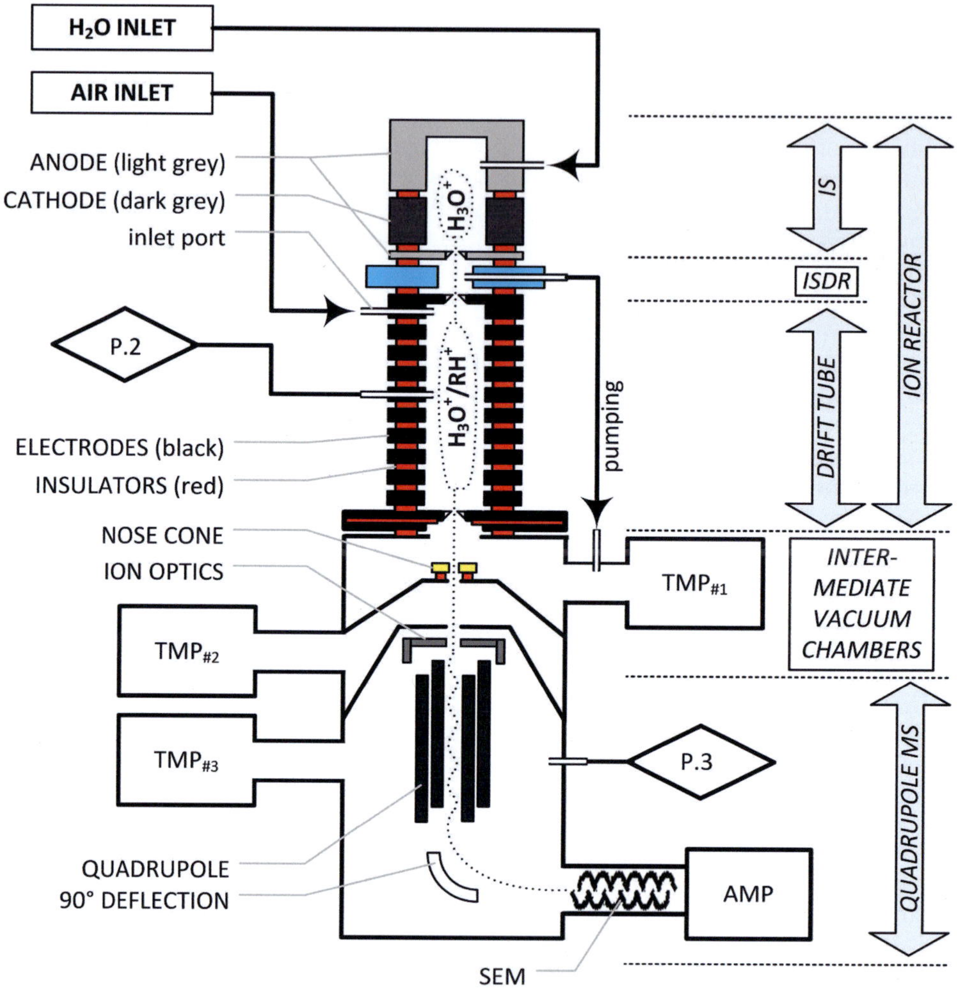

Figure F 2.1: *Schematic diagram of the PTR mass spectrometer with the main components ION REACTOR (with ion source (**IS**), ion source drift region (**ISDR**) and drift tube (**DT**)), INTERMEDIATE VACUUM CHAMBERS and QUADRUPOLE MASS SPECTROMETER (with quadrupole rod configuration, detector (secondary electron multiplier (**SEM**) and signal amplifier (**AMP**)). Other highlighted system components are electrodes (anode, cathode, annular electrodes, nose cone plate, ion optics, 90° deflector), inlet ports (for air and H_2O), pressure sensors **P.x** and the three turbo molecular pumps (**TMP**$_{\#1,\#2,\#3}$).*

2.2 Physics & Chemistry in PTR systems

2.2.1 Ion Generation

Ionization process by electron impact Hydronium is used as primary ion for the ionization of trace gases inside the drift reaction zone. It is produced in a hollow cathode electrical discharge (negative glow) between two anodes (same potential) and the intermediate cathode from water vapor, which is introduced into the IS at a low flow rate of typically $F_{H_2O} = 7.5\,sccm$. Between the nearly field free negative glow region, with a potential close to the anodes on the one side and the cathode on the other side, the disruptive discharge occurs in consequence of the rapidly dropping potential. Hereby mainly energetic electrons collide with neutrals forming the ion products H_2O^+, H_2^+, H^+, O^+ and OH^+, that again undergo further reactions to secondary and tertiary products. The fundamental ion/molecule reactions of ions produced in this source are presented in table T 2.1 (p. 15) together with their specific reaction rate constants k.

Reactions				Rate constants k $[10^{-9}\,(cm^3 \cdot s^{-1})]$
O^+	$+ H_2O$	$\rightarrow H_2O^+$	$+ O$	2.6
H^+	$+ H_2O$	$\rightarrow H_2O^+$	$+ H$	8.2
H_2^+	$+ H_2O$	$\rightarrow \boxed{H_3O^+}$	$+ H$	3.4
		$\rightarrow H_2O^+$	$+ H_2$	3.7
OH^+	$+ H_2O$	$\rightarrow \boxed{H_3O^+}$	$+ O$	1.3
		$\rightarrow H_2O^+$	$+ OH$	1.8
H_2O^+	$+ H_2O$	$\rightarrow \boxed{H_3O^+}$	$+ OH$	1.8

Table T 2.1: *Reactions of ions produced by electron impact upon water vapor with water vapor and corresponding rate constants k. (adapted from Hansel* et al. *[1995])*

Plasma - Ion beam The ions O^+, H^+, H_2^+, H_2O^+ and OH are radicals and therefore highly reactive with correspondingly short chemical lifetimes enabling the final transition of water vapor to H_3O^+ ions within reasonable times. With that the listed ion reactions are appropriate to retain a relatively high density plasma, that finally provides a H_3O^+ ion beam with a purity of at least $99.5\,\%$ [Hansel *et al.*, 1995; Lindinger *et al.*, 1998] into the reaction zone (DT). The final combinations towards almost entirely pure H_3O^+ ions prior to this injection take place inside of the intermediate ion source drift region (ISDR) (see figure F 2.1, p. 14). When using standard values for pressures $P_{ISDR} \sim 1.5\,hPa$, temperatures $T_{ISDR} \sim 313\,K$ and the applied electric field in the ISDR, typical residence times t_{ISDR} are calculated with $\lesssim 10\,\mu s$. Impurities in this ion beam are traces of NO^+ ions, which are produced from N_2 due to back-streaming air into the ISDR, and, as only significant contribution (less than $0.5\,\%$), of O_2^+ ions, that are produced as by-products in the ISDR due to charge-transfers from H_2O^+ ions to O_2 effusing from the DT towards the IS or by direct electron impact ionization of O_2. Both impurity ions can react with trace gases and enhance the background signals of respective masses. Therefore, these ions should be reduced as far as possible. N_2^+ or N^+ rapidly convert to H_3O^+ by multi collisions with H_2O molecules in the DT. [Hansel *et al.*, 1995; Hartungen, 2005; Slevin & Harrison, 1975]

For the final processing of measurement data in terms of accurate quantification of trace gases the normalization to the source output (count rates of H_3O^+ ions) is necessary, and thus primary ion signals are recorded. Typical primary ion rates from hollow cathode discharge IS are in the range of $10^{12} - 10^{13}$ counts per second (cps). However, since the transmission of ions through the instruments is $\ll 1$, the finally observed primary ion rates at the detector (used for normalization) are in the range of (tens of) millions cps.

2.2.2 Proton Transfer - Ion/Molecule Reactions

Proton Transfer Proton transfer, which is the transfer of H^+-ions from one reactant to another, directly depends on the proton affinities (PA) of the participating donor and acceptor molecules, that describe the binding energies of protons (normally given in $kJ \cdot mol^{-1}$ or eV; conversion: $96.485\,kJ \cdot mol^{-1} \approx 1\,eV$) to its atoms or molecules, respectively, and decide whether the reaction is exothermic or endothermic. The chemical transfer reactions basically follow the reaction:

$$\mathrm{XH^+ + R \xrightarrow{k_1} RH^+ + X} \qquad \text{(R 2.13)}$$

with XH^+ being the ion donor and R acting as some proton accepting compound. k_1 depicts the rate coefficient of that reaction. The reaction is exothermic (spontaneous), if PA(R) is higher than PA(X):

$$\Delta PA = PA(R) - PA(H_2O) \qquad \text{(E 2.1)}$$

This is in accordance with the equivalent consideration of the fundamental change in Gibbs free energy ΔG:

$$\Delta G = \Delta H - T \cdot \Delta S \qquad \text{(E 2.2)}$$

where ΔH is the change in free enthalpy and ΔS the change in entropy. The gas phase basicity B, which depicts the willingness of the substance to accept a proton, is the negative of the Gibbs free energy change ΔG for proton transfer. The relation of B and PA is given by [Ellis & Mayhew, 2014]:

$$PA = B - T \cdot \Delta S \qquad \text{(E 2.3)}$$

with PA being defined as the negative of the enthalpy change ΔH. Since ΔS is negligibly small for proton transfers (no crucial change of the net system entropy is assumed), ΔPA determines the spontaneity of a PTR.

Assuming the presented standard case of protonated water (H_3O^+) as ion donor, the proton-transfer reaction leads to equation:

$$H_3O^+ + R \xrightarrow{k_1} RH^+ + H_2O \qquad \text{(R 2.14)}$$

that only takes place, when PA(R) is higher than that of water ($PA(H_2O) = 691\, kJ \cdot mol^{-1}$; NIST Chemistry WebBook[8]). This applies in particular to most of the organic trace gas molecules. Common constituents of air, such as the main components N_2, O_2, Ar, CO_2 and inorganic trace gases, have a lower PA than that of water and therefore do not interfere with the measurement. This wide variety of undetectable compounds in ambient air make the air itself acting as buffer or carrier gas, respectively, that is transparent for proton transferring instruments. An exemplary compilation of proton affinities from relevant atmospheric constituents is given in table T 2.2 (p. 18). More comprehensive data of gas-phase ion energetics are available from NIST Chemistry WebBook or in Hunter & Lias [1998].

The high exothermicity due to a large ΔPA for most VOCs implies, that back reactions are extremely slow, and thus, equilibrium overwhelmingly favors the protonated VOC. However, for molecules with PA(R) only slightly above $PA(H_2O)$, the forward reaction's exothermicity of the ionization process is weak and the back reactions (with rate coefficient k_2):

$$RH^+ + H_2O \xrightarrow{k_2} H_3O^+ + R \qquad \text{(R 2.15)}$$

are not negligible. Although being normally smaller by up to several orders of magnitude, the rate coefficients of the reverse reactions can become significant. This in particular applies in consequence of variations in the sample humidities. Since the amount of water vapor in the DT is normally much higher than the amount of gas R, the forward and backward net reaction rates might get comparable. In that case the overall production of protonated gas R is much less efficient than for other compounds with a larger ΔPA and particularly depends on the sample humidity.

Furthermore, the rate coefficients of the reverse reactions are more temperature dependent and increase with raising temperatures (according Chatelier's principle) [Hansel *et al.*, 1997]. As electric fields are applied to the reaction zone, the effective ion temperatures are increased (see equation E 2.10). This finally enhances the reverse reactions while slightly reducing the forward reactions. [Blake *et al.*, 2009; de Gouw & Warneke, 2007]

8 NIST Chemistry WebBook, NIST Standard Reference Database Number 69; National Institute of Standards and Technology (NIST) [Linstrom & Mallard, 2001, 2011]; available at: `http://webbook.nist.gov/chemistry/`

Type Of Compound	General Chemical Formula
acylperoxy nitrate	R-C(O)OONO$_2$
alcohol	R-OH
aldehyde	R-CHO
alkene	R$_1$C=CR$_2$
amine	R-NH$_2$
aromatics	C$_6$R$_6$ (cyclic)
ketone	RCOR
nitrile	R-C≡N
organic acid . .	R-COOH

Compound	Chemical Sum Formula	Proton Affinity [$kJ \cdot mol^{-1}$]	Protonated Mass [amu]	VOC Class
Helium	He	177.8		
Neon	Ne	198.8		
Argon	Ar	369.2		
Oxygen	O_2	421.0		
Hydrogen	H_2	422.3		
Krypton	Kr	424.6		
Nitrogen	N_2	493.8		
Xenon	Xe	499.6		
Nitrogen Oxide	NO	531.8		
Carbon Dioxide	CO_2	540.5		
Methane	CH_4	543.5		
Nitrous Oxide	N_2O	549.8, 575.2		
Nitrogen Dioxide	NO_2	591.0		
Carbon Monoxide	CO	594.0, 426.3		
Ozone	O_3	625.5		
Sulfur Dioxide	O_2S	672.3		
Water	**H_2O**	**691.0**	**19**	
Hydrogen Sulfide	H_2S	705.0	35	(non-VOC)
Hydrogen Cyanide	HCN	712.9	28	nitrile
Formaldehyde	CH_2O	712.9	31	aldehyde
Formic Acid	CH_2O_2	742.0	47	organic acid
Benzene	C_6H_6	750.4	79	aromatic
Propylene	C_3H_6	751.6	43	alkene
Methanol	CH_4O	754.3	33	alcohol
Acetaldehyde	C_2H_4O	768.5	45	aldehyde
Acetonitrile	C_2H_3N	779.2	42	nitrile
1,3-Butadiene	C_4H_6	783.4	55	alkene
Acetic Acid	$C_2H_4O_2$	783.7	61	organic acid
Toluene	C_7H_8	784.0	93	aromatic
o-Xylene/p-Xylene	C_8H_{10}	796.0/794.4	107	aromatic
Furan	C_4H_4O	803.4	69	enol ether
Acetone/Propanal	C_3H_6O	812.0/786.0	59	ketone/aldehyde
Isoprene	C_5H_8	826.4	69	alkene
MEK	C_4H_8O	827.3	73	ketone
DMS	C_2H_6S	830.9	63	organosulfur compounds
MVK/MACR	C_4H_6O	834.7/808.7	71	ketone
Styrene	C_8H_8	839.5	105	aromatic
Methylamine	CH_5N	899.0	32	amine
PAN	$C_2H_3NO_5$		77	acylperoxy nitrate
Glyoxal	$C_2H_2O_2$		59	aldehyde
α-pinene	$C_{10}H_{16}$		137	monoterpene
C_9-aromatics	C_9H_{12}		121	aromatics
C_{10}-aromatics	$C_{10}H_{14}$		135	aromatics
C_{11}-aromatics	$C_{11}H_{16}$		149	aromatics
C_{12}-aromatics	$C_{12}H_{18}$		163	aromatics
C_{13}-aromatics	$C_{13}H_{20}$		177	aromatics

Table T 2.2: *Components found in ambient air with chemical formula, proton affinities (if clearly determined), protonated mass and classification according fundamental chemical categories with R being the VOC. (adapted from NIST Chemistry WebBook, IONICON, Ellis & Mayhew [2014]; Hunter & Lias [1998]; Seinfeld & Pandis [2006])*

Detailed investigations of the competing forward and backward processes under special attention to sample humidities have been performed for diverse compounds, like e.g. for formaldehyde ($PA(CH_2O) = 712.9\, kJ \cdot mol^{-1}$ [Linstrom & Mallard, 2011]) in Vlasenko *et al.* [2010]; Warneke *et al.* [2011b] and for hydrogen sulfide ($PA(H_2S) = 705\, kJ \cdot mol^{-1}$ [Linstrom & Mallard, 2011]) in Li *et al.* [2014].

Fragmentation/Clustering The chemical ionization via proton transfer is a soft ionization technique without extensive fragmentation of the product ions. However, fragmentation of the product ion RH^+ is observed in specific cases (e.g. for $PANH^+$ [Hansel & Wisthaler, 2000]) and has to be taken into account.

Opposite to fragmentation, namely clustering of hydronium with other water molecules, can occur according to reaction:

$$\mathrm{H_3O^+ + n \cdot H_2O \rightleftharpoons H_3O^+ \cdot (H_2O)_n} \tag{R 2.16}$$

with $n \geq 1$ being the number of involved H_2O molecules. These clusters can also react with gas R and open thus the possibility for proton transfers with higher-order clustered ion donors than just with the initial H_3O^+. The respective equation is:

$$\mathrm{H_3O^+ \cdot (H_2O)_n + R \longrightarrow RH^+ + (H_2O)_{(n+1)}} \tag{R 2.17}$$

with $n \geq 1$. Parent H_3O^+ ions and their water clusters are in principle comparable in their efficiencies of proton transfer reactions with most organic trace gases. But having different raising proton affinities for each $(H_2O)_{1,2,..}$, substances with low PA, that might energetically react only with first members of an ion-family (i.e. only parent ion), are strongly humidity dependent in its sensitivity as consequence of different concentrations of the ion donors.

Another reaction channel of hydrated hydronium clusters is direct switching of ligands, that additionally increases the complexity of data analysis.

Likewise, also product ions can cluster with water molecules according:

$$\mathrm{RH^+ + n \cdot H_2O \rightleftharpoons RH^+ \cdot (H_2O)_n} \tag{R 2.18}$$

with $n \geq 1$ or even directly with gas R according:

$$\mathrm{RH^+ + R \rightleftharpoons RH^+ \cdot R} \tag{R 2.19}$$

Formed cluster ions can then again further react or directly interfere with other species at the same higher masses, which might cause ambiguities concerning the allocation of measurement results. Cluster ion formation and the subsequent reactions of clusters (including proton transfer and ligand switching) do complicate the interpretation of measurements and should be avoided as far as possible (by proper control of electric fields in the different instrument compartments).

2.2.3 Ion Kinetics in the Drift Tube Reactor

Inside the drift tube (reaction chamber) the trace gases in the sampled air undergo mostly non-dissociative proton transfer from hydronium (see section 2.2.2, p. 16). In order to guide the ions toward the exit of the DT, an electrical field is applied across the DT reaction zone, which accelerates the ions before their extraction to the mass spectrometer. By this electrical forcing the ions gain a drift velocity ν_{drift}, which can be expressed as:

$$\nu_{drift} = \mu \cdot E \quad \text{(E 2.4)}$$

$$= \mu_0 \cdot \frac{P_0}{P} \cdot \frac{T}{T_0} \cdot E \quad \text{(E 2.5)}$$

$$= \mu_0 \cdot N_0 \cdot (\frac{E}{N}) \quad \text{(E 2.6)}$$

with:

ν_{drift}:	drift velocity	$[cm \cdot s^{-1}]$
μ/μ_0:	ion mobility / standard value	$[cm^2 \cdot V^{-1} \cdot s^{-1}]$
E:	electrical field	$[V \cdot cm^{-1}]$
P/P_0:	pressure / standard value ($1\ atm$)	$[hPa]$
T/T_0:	temperature / standard value (273.15)	$[K]$
N/N_0:	gas number density / standard value at p_0, T_0	$[cm^{-3}]$

with μ being the ion mobility, E the electrical field, P the pressure, T the temperature and N the gas number density. Zero-labels indicate standard conditions. The ion mobility μ has been determined for numerous ions in different buffer gases, including H_3O^+ ions in nitrogen [Dotan *et al.*, 1976; Viehland & Mason, 1995]. Following equations E 2.4 - E 2.6 in consideration of the ideal gas law, the ion velocity can be expressed using the ratio $^E/_N$ as reduced electric field parameterization, which is oftentimes introduced to describe ion mobility. The corresponding unit is Townsend ($1\ Td = 10^{-17}\ V \cdot cm^2$). Extensive tabulations of ionic energy parameters covering diverse gaseous ions and reduced mobility have been published in Ellis *et al.* [1978, 1984, 1976] and were discussed with particular connection to PTR-MS techniques in Ellis & Mayhew [2014].

Ion kinetic energies, Ion temperatures The selective heating of the ions induced by the electric field E causes $T_{ion} > T_{drift}$ and with that an average center-of-mass kinetic energy of ion collisions with neutrals in the buffer gas $\langle KE_{center-of-mass} \rangle$ (associated with temperature $T_{collision}$) of (using Wannier expressions [de Gouw *et al.*, 2003; Viggiano & Morris, 1996]):

$$\langle KE_{center-of-mass} \rangle = \frac{1}{2} \cdot \frac{(m_{ion} + m_{buffer}) \cdot m_{neutral}}{(m_{ion} + m_{neutral})} \cdot \nu_{drift}^2 + \frac{3}{2} \cdot k_B \cdot T_{drift} \quad \text{(E 2.7)}$$

$$= \frac{3}{2} \cdot k_B \cdot T_{collision} \quad \text{(E 2.8)}$$

with:

m_{ion}:	mass of ions	
m_{buffer}:	mass of buffer gas	
$m_{neutral}$:	mass of neutrals	
k_B:	Boltzmann constant	$k_B = 1.3806488(13) \times 10^{-23}\, J \cdot K^{-1}$ [Mohr *et al.*, 2012]

$T_{collision}$ is obviously individual for every different neutral compound in the DT. For collisions with the buffer gas itself (i.e. air) ($m_{buffer} = m_{neutral}$), equation E 2.8 transforms to:

$$\langle KE_{center-of-mass}\rangle_{buffer} = \frac{1}{2} \cdot m_{buffer} \cdot \nu_{drift}^2 + \frac{3}{2} \cdot k_B \cdot T \qquad \text{(E 2.9)}$$

$$= \frac{3}{2} \cdot k_B \cdot T_{eff} \qquad \text{(E 2.10)}$$

with T_{eff} being described as the internal energy of the ion reactants [Viggiano & Morris, 1996].

The reagent ions' kinetic energies (together with internal energies as functions of T) are the regulating parameters concerning formation, clustering and fragmentation of ions. In consideration of the reaction zone pressure (here: $P_{drift} = 2.2\, hPa$), the kinetic energies control/determine the residence times of the ions in the drift zone (typically $t_{drift} \sim 100\, \mu s$ calculated for a reaction zone temperature of $T_{drift} = 313\, K$) and therewith the collision rates with neutrals (i.e. sampled trace gases).

- **High kinetic energy (high $^E/_N$):**
  A high kinetic energy of ions leads to a short stay and with that to a reduced collision probability. The result of little collisions is a generally reduced cluster formation and weaker ion beam broadening and thus higher ion transmission rate.

- **Low kinetic energy (low $^E/_N$):**
  A low kinetic energy means longer residence times and with that increased collision frequencies of primaries. The result is enhanced sensitivity.

The applied voltage (together with P and T) therefore governs the entire collision frequencies with appropriate collision energies inside of the drift zone, balancing between lower degree of cluster formation as consequence of faster ions, and higher sensitivity as consequence of slower ions. In order to allow PTR processes on a reasonable level, a complete suppression of appearing clusters is obviously almost impossible.

Nevertheless, a value ideally balancing residence time and ion kinetic energy of primary ions in the drift tube, that clearly reduces the formation of hydronium water clusters, has been found with an $^E/_N \sim 120\, Td$ (typical value range: $100 - 140\, Td$). Lower values increase the sensitivity. Using distinctly higher values further reduces the proportion of hydrated hydronium clusters, but at the same time increases the probability for other unwanted fragmentation effects.

In order to derive the yield fractions of transfers from parent to product ions, the residence time Δt in the drift tube has to be inferred. Under the assumption that only reactions with a single organic trace gas R are possible via the reaction channel of proton transfer, the amount of R in the sampled air that gets ionized by this proton transfer during its sojourn in the drift reactor Δt is given by:

$$[RH^+] = [H_3O^+]_0 \cdot (1 - exp^{(-k \cdot [R] \cdot \Delta t)}) \qquad \text{(E 2.11)}$$
$$\approx [H_3O^+] \cdot k \cdot [R] \cdot \Delta t \qquad \text{(E 2.12)}$$

where k is reaction rate constant for reaction R 2.14 (p. 17). Assuming $[H_3O^+] \gg [RH^+]$, which is applicable and therefore allowed for that simplification, the linear dependency in [R] allows the direct calculation of the drift velocity ν_{drift}:

$$\nu_{drift} = \frac{L_{drift}}{\Delta t} \qquad \text{(E 2.13)}$$

with L_{drift} being the length of the DT (here: typically $L_{drift} = 8.8\,cm$). By using this connection together with the ion velocity from equation E 2.6 (p. 20), the residence time Δt can be expressed as:

$$\Delta t = \frac{L_{drift}}{\mu_0 \cdot N_0} \cdot \left(\frac{E}{N}\right)^{-1} \qquad \text{(E 2.14)}$$

Finally the combination of the derived residence time Δt with the concentration estimation according to equation E 2.12 allows the calculation of the yield fraction of the transfer from parent to product ions:

$$\frac{[RH^+]}{[H_3O^+]} = [R] \cdot \frac{k \cdot L_{drift}}{\mu_0 \cdot N_0} \cdot \left(\frac{E}{N}\right)^{-1} \qquad \text{(E 2.15)}$$
$$= VMR \cdot \frac{k \cdot L_{drift}}{\mu_0 \cdot N_0} \cdot \frac{N^2}{E} \qquad \text{(E 2.16)}$$

with the volume mixing ratio $VMR = {}^{[R]}/_{N}$ of gas R.

The fractions of the particular ion concentrations [RH^+] and [H_3O^+], respectively, or more specifically of the corresponding count rates i(H_3O^+) and i(RH^+), respectively, are detected by the QMS and allow the calculation of the ratio ${}^{i(RH^+)}/_{i(H_3O^+)}$. Therefore the mass-individual transmission efficiencies with associated transmission factors $T_{H_3O^+}$ and T_{RH^+} have to be determined for the particular system and parameters, because they strongly depend on individual conditions and settings also influencing extraction and detection, like e.g. pressures and electric potentials.

A fraction of the primary and product ions is sampled into the QMS. The count rates observed at the detector are related to the quantity to be determined, the VMR, by:

$$\frac{i(RH^+)}{i(H_3O^+)} = \frac{[RH^+]}{[H_3O^+]} \cdot \frac{T_{RH^+}}{T_{H_3O^+}} \tag{E 2.17}$$

$$= VMR \cdot \frac{k \cdot L_{drift}}{\mu_0 \cdot N_0} \cdot \frac{N^2}{E} \cdot \frac{T_{RH^+}}{T_{H_3O^+}} \tag{E 2.18}$$

In order to enable the comparison of PTR instruments as well as measurements, the particular count rates are normalized to the number of detected primary ions. Therefore it is common practice to normalize the measuring data to a H_3O^+ signal of one million counts per second ($10^6\,cps$). A VMR of 1 part per billion by volume ($ppbV$: 10^{-9}) of a trace gas R gives then a sensitivity ε_R of:

$$\varepsilon_R = 10^{-3} \cdot \frac{k \cdot L_{drift}}{\mu_0 \cdot N_0} \cdot \frac{N^2}{E} \cdot \frac{T_{RH^+}}{T_{H_3O^+}} \tag{E 2.19}$$

by default using units of normalized counts per second per $ppbV$: [$^{ncps}/_{ppbV}$], which is typically in the range of $10 - 40\,^{ncps}/_{ppbV}$ for a broad variety of VOCs with $50 - 1000\,^{cps}/_{ppbV}$ (see chapter 4, p. 57). The instrumental sensitivity ε is equivalent to the compound's individual calibration factor S.

As consequence for humidity dependent effects during proton transfer (discussed in section 2.2.2, p. 16) the normalization has to be performed in consideration of water clustering. This basically means, that the normalization of $i(RH^+)$ by $i(H_3O^+)$ according equation E 2.18 (p. 23) has to be extended to (only clusters for $n = 1$ are to be considered $\leftrightarrow$ reaction R 2.16, p. 19):

$$\frac{i(RH^+)}{i(H_3O^+)} \Longrightarrow \frac{i(RH^+)}{i(H_3O^+) + X_R \times i(H_3O^+(H_2O))} \tag{E 2.20}$$

The factors X_R are specific for each compound and have to be determined empirically by using gas standards with different humidities [de Gouw *et al.*, 2003; de Gouw, 2003]. In general, many compounds showed X_R factors of ~ 0.5, but aromatics have clearly lower values [Blake *et al.*, 2009; Warneke *et al.*, 2001].

Accuracy According equation E 2.19 (p. 23) the systems' accuracies are theoretically limited by the uncertainties of the rate constant k (assumed to be $\leq 50\,\%$) and the transmission ratios $^{T_{H_3O^+}}/_{T_{RH^+}}$ (assumed to be $\leq 25\,\%$) [de Gouw & Warneke, 2007]. The comparison between calculated and measured sensitivities effectively showed differences of up to a factor of 2 [Warneke *et al.*, 2003], which demands for an enhancement when doing atmospheric measurements. A good estimate for average accuracy is limited to approximately $\sim 20 - 25\,\%$ for PTR measurements, when using predefined mixtures of VOCs as gas calibration standards [Blake *et al.*, 2009; de Gouw & Warneke, 2007].

Precision The instrument precision is mostly limited by the ion counting statistics of the product ions, which are expected to be represented by Poissonian distributions [Hayward *et al.*, 2002]. Therefore the instrument noise can be assumed to be proportional to the square-root of the signal and the random scatter can be described using the standard deviation σ [Ellis & Mayhew, 2014]:

$$\sigma = \frac{\sqrt{N}}{\tau} \tag{E 2.21}$$

$$= \sqrt{\frac{i}{\tau}} \tag{E 2.22}$$

with N being the total number of counts, τ the measurement time and i the count rate in $[cps]$.

Assuming a VMR_R of $1\,ppbV$ for compound R with $\varepsilon_R = 25\,^{ncps}/_{ppbV}$ and a primary ion rate of $i(H_3O^+) = 2 \times 10^7\,cps$, the instrumental precision then gives (in $1\,s$):

$$\frac{\sqrt{N}}{N} = \frac{\sqrt{500}}{500} \tag{E 2.23}$$

$$\cong 4.5\% \tag{E 2.24}$$

which is only due to statistical reasons. This clearly demonstrates that PTR measurements in terms of instrumental precision improve when mixing ratios or signals, respectively, increase:

$$\Delta N \propto \frac{1}{N} \tag{E 2.25}$$

Likewise longer dwell times τ enhance the precision of measurements as well.

Limit of Detection (LOD) In order to derive the LOD (general principles discussed in Currie [1968]), noise by means of the instrumental background has to be determined through direct comparison of ambient (external + instrumental) and background (only instrumental) contributions to the measurement signals. The LOD can then be defined as the mixing ratio of compound R with a signal-to-noise ratio (S/N) of 3 ($^{\Delta VMR}/_{VMR} = 33\,\%$) [de Gouw & Warneke, 2007] in order to reliably distinguish a signal from the background noise [Ellis & Mayhew, 2014]:

$$LOD_R = \frac{3 \times VMR_R}{(S/N)} \tag{E 2.26}$$

$$= \frac{3 \times \sigma_{background}}{\varepsilon_R} \tag{E 2.27}$$

where $\sigma_{background}$ is the standard deviation of the background signal (background conditions; S: signal level; N: noise level) and ε_R the appropriate instrumental sensitivity. The limit of detection is typically in the range of $50 - 150\,pptV$ for measurements with integration times of $1\ s$, and therefore sufficient for atmospheric high-sensitivity trace gas measurements. The explicit calculation of VMR and its error ΔVMR can be expressed according [de Gouw & Warneke, 2007; de Gouw, 2003]:

$$VMR = \frac{1}{I_{H_3O^+} \times S}(I_{ambient} - I_{background}) \tag{E 2.28}$$

$$\Delta VMR = \frac{1}{I_{H_3O^+} \times S}\sqrt{\frac{I_{ambient}}{\tau_{ambient}} + \frac{I_{background}}{\tau_{background}}} \tag{E 2.29}$$

with:

VMR:	volume mixing ratio	$[ppbV]$
$I_{H_3O^+}$:	count rate of H_3O^+	$[10^6\ counts\cdot sec^{-1}]$
S:	sensitivity/calibration factor	$[ncps\cdot ppbV^{-1}]$
$I_{ambient}$:	count rate of ions with catalyst off	$[cps]$
$I_{background}$:	count rate of ions with catalyst on	$[cps]$
$\tau_{ambient}$:	ambient dwell time	$[s]$
$\tau_{background}$:	background dwell time	$[s]$

The applied equations describe the precision of the measurement as function of the VOC mixing ratios. Relative errors, gradually increasing with decreasing mixing ratios, can be $> 100\%$ for very low ratios. The LOD improves with increasing dwell times.

2.3 Quadrupole Mass Filtering

For the mass analyzing detection system in PTR-MS instruments oftentimes quadrupole mass spectrometers (QMS) are used. These QMS devices offer a good balance between measurement performance and instrumental complexity and are therefore widely engaged in standard mass spectrometric tasks. Nonetheless, it has to be mentioned again that advancements in instrument developments towards time-of-flight and ion-trapping setups provide new concepts that rival the solely use of quadrupole mass spectrometers, but due to its preferences do not question but complement its applications.

Concept of QMS The basic idea of a QMS comprises mass filtering of ions inside a high-frequency quadrupole field and subsequent detection of then preselected ions via e.g. secondary electron multiplier (SEM). The filtering and detection system is kept under high-vacuum with pressures $P_{QMS} < 3 \times 10^{-5}\, hPa$. This is required in order to reduce the probability of unwanted collisions with air molecules by maximization of the mean free path of ions during the selection process. At the same time the aging of the SEM due to oxidation is minimized and electrical flashovers are avoided. The ions to be probed are injected into this high-vacuum chamber through a system of orifices and focusing einzel lenses[9] along stepwise reduced pressure stages.

The mass selection itself is achieved by using a quadrupole mass filter, which filters ions according to individual mass-discriminating properties using their specific mass-to-charge ratio (m/z). The filter consists of 4 metal rods (here: molybdenum) with an ideally hyperbolic (or circular) cross section serving as electrodes with direct connections between opposing rods, that are symmetrically and equidistantly arranged around a center axis z with radius r_0 in a radial array. The particular arrangement is depicted in figure F 2.2 (p. 27).

Theoretic demonstration of the filtering processes in the QMS Following the mathematical standard deduction of quadrupole filtering as discussed in Douglas [2009], the electric potential of the quadrupole $\phi(x, y)$ is then given by:

$$\phi(x, y) = \phi_0 \cdot \frac{(x^2 - y^2)}{r_0^2} \tag{E 2.30}$$

which satisfies the Laplace's equation. The quadrupole rod pairs are each fed by a constant time-independent direct-current (DC) voltage U and are superimposed by a periodically alternating time-dependent radio frequency (RF) voltage V_{RF} with:

$$V_{RF} = V_0 \cdot cos(\omega t) \tag{E 2.31}$$

[9] Einzel lens: ion optical lenses focusing charged particles

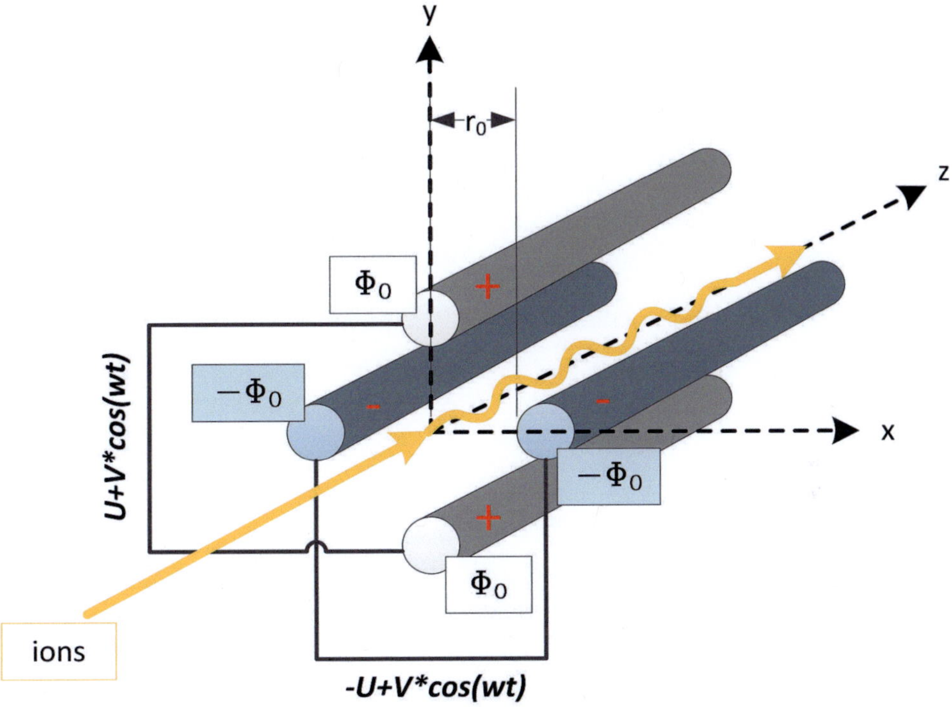

Figure F 2.2: *Schematic diagram of a quadrupole with appropriate voltage supply of the two rod pairs and resulting potentials ϕ. An exemplary ion trajectory is indicated in yellow.*

where V_0 is the amplitude, ω its frequency and t the time, and has a 180 degree phase shift in between. The potentials ϕ_0 that are applied to the quadrupole rods are given by:

$$\phi_0(t) = U - V_{RF} \cdot cos(\omega t) \tag{E 2.32}$$

The electric potential Φ of the quadrupole can therefore generally be described as:

$$\Phi(x, y, t) = \phi_0(t) \cdot \frac{(x^2 - y^2)}{r_0^2} \tag{E 2.33}$$

The trajectories of the traversing ions consist of spiral tracks along the longitudinal axis of the system. Focusing and defocusing effects in x- and y-directions as consequence of applied frequencies ω finally allow the finding of matching parameters for stable trajectories through the field potentials. In order to describe the ion trajectories in that field, Newton's law of motion can be used with:

$$\vec{f} = m \cdot \frac{d\vec{v}}{dt} \tag{E 2.34}$$

where $\vec{f}$ is the force on the ion with mass m, velocity $\vec{v}$ and time t [Douglas, 2009]. A positive ion is exposed to a force of:

$$\vec{f} = -q \cdot \vec{\nabla}\Phi(x, y, t) \tag{E 2.35}$$

where q is the charge of the ion with z being the number of elementary charges e on the ion ($q = z \cdot e$). The equilibrium of forces:

$$m \cdot \frac{d\vec{v}}{dt} = -q \cdot \vec{\nabla}\Phi(x, y, t) \tag{E 2.36}$$

then allows to derive the equations of motion in the respective directions x,y,z:

$$\frac{d^2x}{dt^2} = -\frac{2qx}{mr_0^2}\phi_0(t) \tag{E 2.37}$$

$$\frac{d^2y}{dt^2} = +\frac{2qy}{mr_0^2}\phi_0(t) \tag{E 2.38}$$

$$\frac{d^2z}{dt^2} = 0 \tag{E 2.39}$$

The electric fields $E_{x,y,z}$ in the corresponding directions x, y and z are therefore [Douglas, 2009]:

$$E_x = -\frac{\partial\Phi(x, y, t)}{\partial x} \tag{E 2.40}$$

$$= -\frac{2x}{r_0^2} \cdot \phi_0(t) \tag{E 2.41}$$

$$E_y = -\frac{\partial\Phi(x, y, t)}{\partial x} \tag{E 2.42}$$

$$= -\frac{2y}{r_0^2} \cdot \phi_0(t) \tag{E 2.43}$$

$$E_z = -\frac{\partial\Phi(x, y, t)}{\partial z} \tag{E 2.44}$$

$$= 0 \tag{E 2.45}$$

With $a_{x,y}$, $b_{x,y}$ and ξ used according the following standard transformations [Douglas, 2009; Sprung, 2002]:

$$a_x = -a_y = \frac{8qU}{m\omega^2 r_0^2} \tag{E 2.46}$$

$$b_x = -b_y = \frac{4qV_{RF}}{m\omega^2 r_0^2} \tag{E 2.47}$$

$$2\xi = \omega t \tag{E 2.48}$$

the equations of motion for the ion in the quadrupole field can be expressed as:

$$\frac{d^2u}{d\xi^2} + (a_u - 2b_u cos2\xi)u = 0 \tag{E 2.49}$$

with $u = x, y$. This equation E 2.49 (p. 29) describes the ion trajectories using parameters a_u from equation E 2.46 (p. 28) and b_u from equation E 2.47 (p. 29), that represent the input parameters DC voltage (U), RF voltage (V_{RF}), radius (r_0), frequency (ω), charge (q) and mass (m) of ions in the quadrupole field. It allows the operation of the quadrupole mass filter to be reduced from a six-dimensional problem to a two-dimensional problem [Miller & Denton, 1986]. The presented mathematical equation is also known as the linear ordinary differential equation of Mathieu (discussed in detail in the NIST Handbook of Mathematical Functions [Olver *et al.*, 2010]). Its solutions are oscillations, whose amplitudes either remain finite (bounded solutions) or grow exponentially (unbounded solutions).

Stability diagram The corresponding solution functions $\beta_u(a_u, b_u)$ spread out the stability diagram in figure F 2.3 (p. 30), that shows the resulting stable region where the amplitudes are smaller than r_0. The used peak of the stability triangle is located at $a_0 = 0.23699$ and $b_0 = 0.706$ [Dawson, 1976]. Ions with mass m and charge q are allowed to pass the quadrupole when their corresponding values a_u and b_u are within this stability region. All others being outside the shaded area potentially collide with the rods and are neutralized. While using $U = 0$ or $a_u = 0$, respectively, all ion trajectories with $b_u < 0.908$ are stable and represent the integral mode of the quadrupole mass filter. The filtering process in general can therefore be understood as a combination of band-pass mass filters.

For constant ratios of U and V_{RF}, the respective mass-to-charge ratios are located on the mass scan line through the origin, that is exemplarily indicated by labeled circles with corresponding ion weights $m1 < m2 < m3$ in figure F 2.3 (p. 30). In order to prevent different m/q matching the stable conditions, the mass scan line is preferentially chosen such that only one m/q fits in the stability region and can pass the quadrupole filter. The corresponding ratio of a_u (from equation E 2.46, p. 28) and b_u (from equation E 2.47, p. 29) ($u = x, y$) is:

$$\frac{a_u}{b_u} = 2 \cdot \frac{U}{V_{RF}} \tag{E 2.50}$$

and with that the ratio of the voltages U and V_{RF} is:

$$\frac{U}{V_{RF}} = 0.168 \tag{E 2.51}$$

which represents the slope of the mass scan line of the quadrupole mass filter. Constant ratios U/V_{RF} imply constant resolution across the m/q spectrum, but it is more common to operate the quadrupole such, that constant peak width is enabled across the spectrum and with that a unit resolution. This requires slightly non-linear mass scan lines [Douglas, 2009].

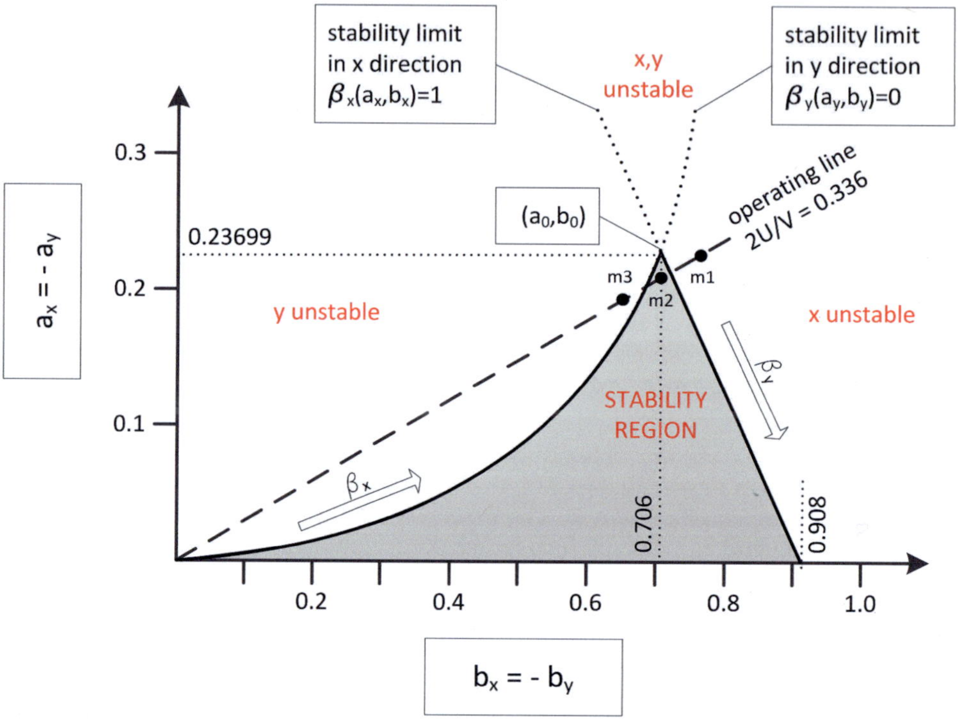

Figure F 2.3: *Stability diagram of a quadrupole mass filter indicating (un-)stable solution areas of the mathieu equations in x,y directions for the first stability region. A typical operating line with masses $m_1 < m_2 < m_3$ is shown. (reproduced from Campana [1980]; Dawson [1976]; Douglas [2009]; Miller & Denton [1986])*

Ion trajectories inside of the QMS The ion injection and following ion movement along the z axis is not a very crucial parameter concerning reasonable results, since the process bases on stable trajectories rather than on a low-divergence ion beam. Nonetheless, a focused beam with low divergence using ion optics (einzel lenses) into the center axis of the quadrupole enhances the performance of the QMS. Used parameters of the lenses and the quadrupole are therefore concerted in order to achieve appropriate ion velocities and with that residence times inside the quadrupole fields allowing the ions to undergo adequate oscillations according the

RF-field properties. Instrumental mass resolution and transmission efficiencies of the quadrupole also depend on this coordination.

QMS in practice In the practical use the stability diagram is mostly not exactly met, since deviations from the theoretically derived ideal prerequisites, like e.g. deformed boundary fields due to finite lengths of the rods, effects due to mechanical deviations of the setup or field imperfections due to other electro-magnetic influences, do occur and need to be considered.

In the presented setting both DC (U) and RF (V_{RF}) voltages need to be changed each time when probing different masses or m/q, respectively. This limitation of one detected m/q per setting implies essentially a separate detection of particular m/q one after the other. Notwithstanding, due to fast electronics a high detection duty cycle of up to multiple m/q per second is possible. The measurement pattern therefore normally comprises peak hopping from mass to mass, whose corresponding m/q values and with that a_u and b_u are either individually selected or incrementally increased. Likewise, while scanning over U and V together with constant ratio, ions of increasing mass get into the stability region in order of their mass and are transmitted sequentially giving a full mass spectrum. As quadrupoles always filter according to m/z, it has to be considered that a difference between one mass to one charge and two masses to two charges cannot be resolved with such a system. The geometric parameters of the quadrupole, namely length and diameter of the rods in combination with their shapes and distances, also determine the mass range and resolution of such an instrument.

A more comprehensive description of quadrupole mass spectrometry with applications can be found e.g. in Dawson [1976].

Part II

Instrumental:
The Ultra-Light-Weight PTR-MS (ULW-PTR-MS)

The instrument ULW-PTR-MS (acronym for Ultra-Light-Weight Proton-Transfer-Reaction Mass Spectrometer) is a newly developed fast-response high-sensitivity in situ measuring system for ambient air VOCs. The system is intended to be deployed aboard the German research aircraft HALO[10] (acronym for High Altitude and LOng Range) in the first instance during the aircraft mission OMO (see section 5.2, p. 86). Its major advantages are its compact size and light weight, that also reflects the instrument recognition Ultra-Light-Weight. Both attributes are crucial factors when deploying instruments on moving research platforms and oftentimes decide about actual deployment options. With a weight-based financial participation for the use of scientific equipment aboard HALO, these features are additional advantages concerning the financing of such projects and justify the previous expenditures.

The ULW-PTR-MS is oriented on the PTR mass spectrometer aboard the CARIBIC[11] passenger aircraft (acronym for Civil Aircraft for the Regular Investigation of the atmosphere Based on an Instrument Container [Brenninkmeijer *et al.*, 1999]), which has been particularly described by Sprung & Zahn [2010]. This system is regularly in use since 2005 in an automated instrument container installed to a Lufthansa Airbus A340-600 (see also section 5.1, p. 83). The CARIBIC PTR-MS is a strongly modified standard mass spectrometer by IONICON[12] in Innsbruck/Austria. The independent further development towards the lightweight laboratory instrument PTR-IF-MS (acronym for Proton-Transfer-Reaction Mass Spectrometer with a DT experimentally enhanced with an ion funnel (IF)), which has been presented by Brito [2011] in 2011, constitutes another substantial basis for the further development to the ULW-PTR-MS. However, the completion of the developments of PTR-IF-MS towards an instrument capable of flying and with the needed instrumental performance and reliability for atmospheric high-precision measurements aboard HALO had not been fully achieved by the start of this project in 2011 and its aeronautical certification was still pending. Furthermore, the planned HALO mission OMO (originally scheduled for 2009; then continually shifted to 2013) has been postponed once more anyway (currently expected to take place in 2015) (see section 5.2.2, p. 88). In order to make use of this then existing situation in early spring 2011, a complete redesign and therewith associated upgrade in consideration of the aforementioned aspects has been decided. For the construction of the new system the former existing vacuum chambers of PTR-IF-MS have been adopted. The system architecture of ULW-PTR-MS bases on the newly developed electronics with associated software and was first-time presented in mid 2013 in the context of the aeronautical certification process for its deployment aboard HALO (see also section 5.2.1, p. 86). Likewise, the intended double use of the new system in form of a sister model, that will be operated in parallel aboard the CARIBIC aircraft, also pushed these progresses. The planned replacement of the meanwhile aged

10 `http://www.halo.dlr.de/`
11 `http://www.caribic-atmospheric.com/`
12 `http://www.ionicon.com/`

CARIBIC PTR-MS with this updated and customized instrument is expected to take place in spring 2015.

The ULW-PTR-MS performed successfully for the first time in a pre-final development version during the field mission UBWOS 2012 in Utah/USA (see part III, p. 95) and has been comprehensively characterized in that context (see section 4, p. 57).

Instrumental testings during research flights in connection with the MUSICA measurement campaign on the Canary Islands/Spain have been performed in July 2013 using the identical sister instrument of ULW-PTR-MS. Selected instrumental results attesting the airworthiness of the system are attached to appendix D (p. 169).

This part of the presented work introduces to the new instrument ULW-PTR-MS. On this occasion the following chapter 3 (p. 37) gives a technical overview of the instrument and introduces to the features of the different subsystems. The system's characterization (including calibration and performance assessments) is presented in chapter 4 (p. 57). Finally, future airborne applications within the CARIBIC project and aboard HALO are presented in chapter 5 (p. 83).

Chapter 3

Technical Setup

The following system description presents the final configuration level of the ULW-PTR-MS as it is certified during the aviation approval for its deployment aboard HALO and only system specific attributes are highlighted. More detailed basic instructions and references to construction manuals (i.e. mechanical drawings, wiring diagrams of the electric system, technical and safety data sheets, etc.) are compiled in the corresponding official approval documents summarized under its mission identification reference HKMS (HALO - KarlsruheMassSpectrometer), which has been assigned within the scope of the aeronautical certification (see section 5.2.1, p. 86).

3.1 Modular design

Technical guidelines The PTR system is designed to be deployed aboard the HALO aircraft. Clear constraints with respect to dimensions and weight demanded a compact and lightweight system that fulfills technical requirements covering on safety aspects for airborne applications. The corresponding universal guidelines for the use aboard HALO are specifically constituted by the German Aerospace Center[13] (DLR), who operates the aircraft on behalf of the participating partners [DLR, 2009].

Dimensions The instrument has a net weight of $\sim 55\,kg$, which includes the fixation kits for integration into the certified HALO aircraft rack. The used aluminum rack, fitting to the payload bays inside of the aircraft cabin, has a supplementary weight of $\sim 17\,kg$ and is the holding fundament of the setup (aircraft base structure with pigeonhole $+15\,kg$). Despite its safety-related design, the instrument is lightweight compared to commercially available PTR mass spectrometers that typically weigh $> 130\,kg$. Figures F 3.1 (p. 38) and F 3.2 (p. 39) show CAD drawings of the instrument integrated into the HALO rack in front and rear views, that have been made in the context of strength calculations for the system's aeronautical certification.

[13] http://www.dlr.de/

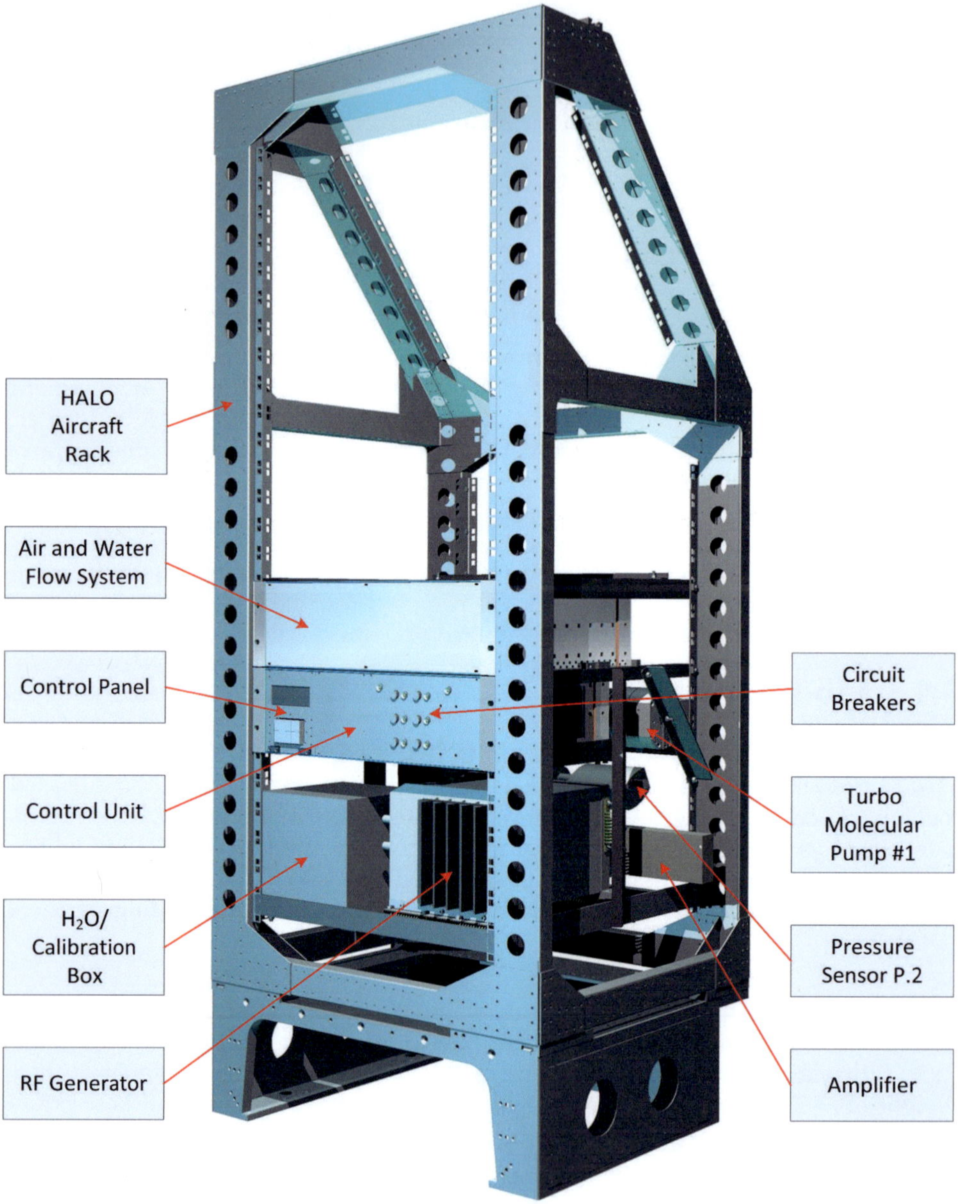

Figure F 3.1: *CAD model: Front perspective of ULW-PTR-MS mounted in a HALO aircraft rack. Diverse aluminum skirts (used as electromagnetic shielding) and protective housings (used as high-voltage contact protections) are transparent in this illustration. Cables and tubes are not shown for reasons of clarity.*

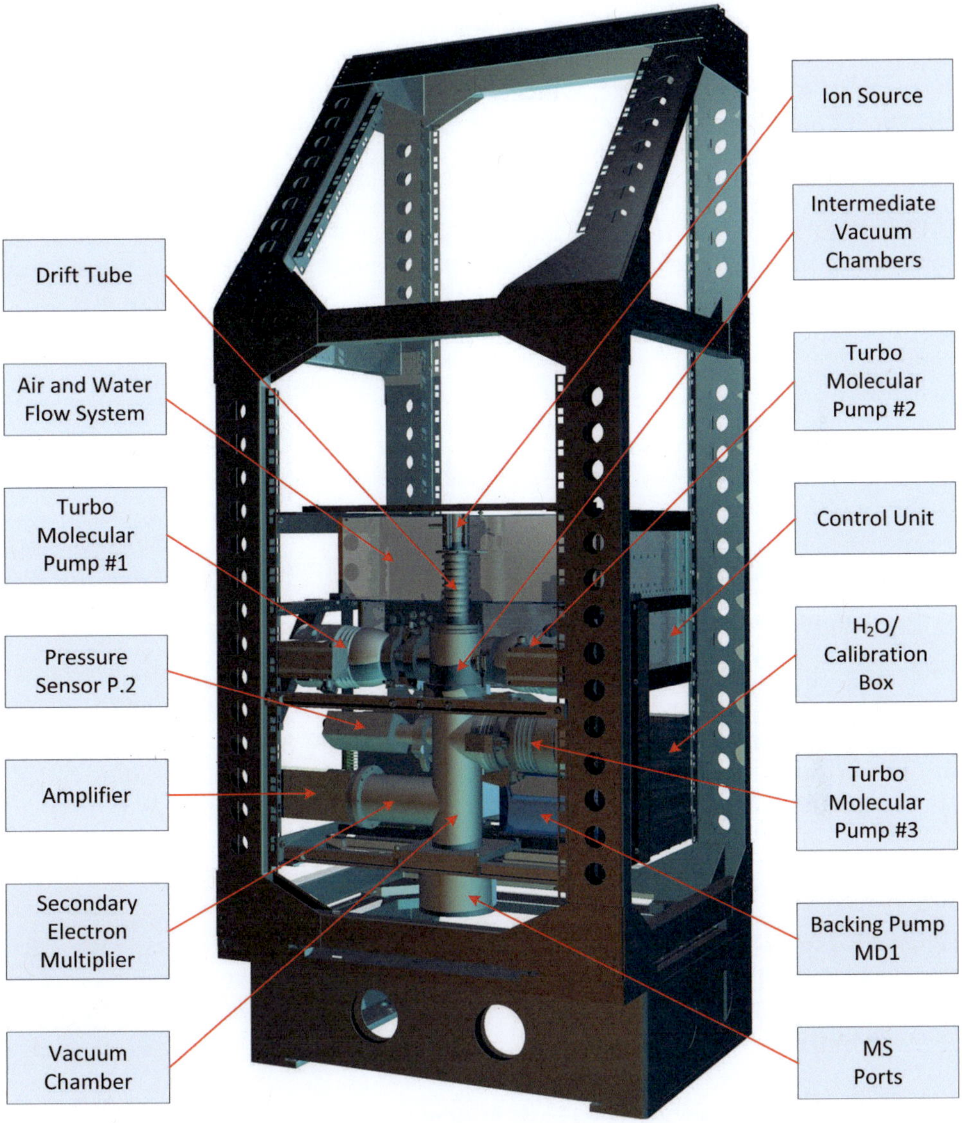

Figure F 3.2: *CAD model: Rear perspective of ULW-PTR-MS mounted in a HALO aircraft rack. Diverse aluminum skirts (used as electromagnetic shielding) and protective housings (used as high-voltage contact protections) are transparent in this illustration. Cables and tubes are not shown for reasons of clarity.*

A picture of ULW-PTR-MS during operations in a test chamber is presented in figure F 5.3 (p. 87). Its low mounting position in the rack has been chosen in accordance with specifications of the licensed rack envelope of max. $150\,kg$ for instrumentation [DLR, 2009]. In this way the center of gravity of the fully loaded rack has been placed as low as possible, which allows the installation of a second smaller instrument into the free upper position of the rack. During the OMO mission the coordinated partner instrument is the ozone device FAIRO [Zahn *et al.*, 2012].

Units With respect to technical aspects like spatial arrangement of functional equipment, maintenance accessibility and certification requirements, the system is constructed in a modular design consisting of six semi-independent assembly units:

1. **Control Unit** - see section 3.1.1 (p. 40)
2. **Air and Water Flow System** - see section 3.1.2 (p. 43)
3. **Ion Reactor** - see section 3.1.3 (p. 46)
4. **Intermediate Vacuum Chambers** - see section 3.1.4 (p. 49)
5. **Quadrupole Mass Spectrometer** - see section 3.1.5 (p. 50)
6. **Assistance Systems** - see section 3.1.6 (p. 51)

The total system's schematic overview shown in figure F 3.3 (p. 41) illustrates the six units in form of shaded areas. Single devices are labeled by their identifier codes. The compilation of used standard equipment is attached to appendix A (p. 159).

3.1.1 Control Unit

Figure F 3.4 (p. 42) shows the Control Unit of ULW-PTR-MS comprising the integrated micro computer V25 with control panel, the internal power supply as well as the power and signal connections to all subsystems and sensors. The unit is being installed into a 19″ subrack.

Micro computer system V25 As basic system of the controls the micro computer V25 is used, that has been developed at the Max Planck Institute for Chemistry (MPIC) in Mainz/Germany [MPIC, 2012]. The individualized construction system of this standalone control unit is integrated in a customized setup complying with the demands of the PTR mass spectrometer. A compilation of used modules is presented in table T 3.1 (p. 43). As the used set consists of only 12 of 16 potential modules by now, that are distributed over the 2 backplanes with 8 slots each (Σ 16 slots in total + extra interfaces), a further expansion with up to 4 units is possible

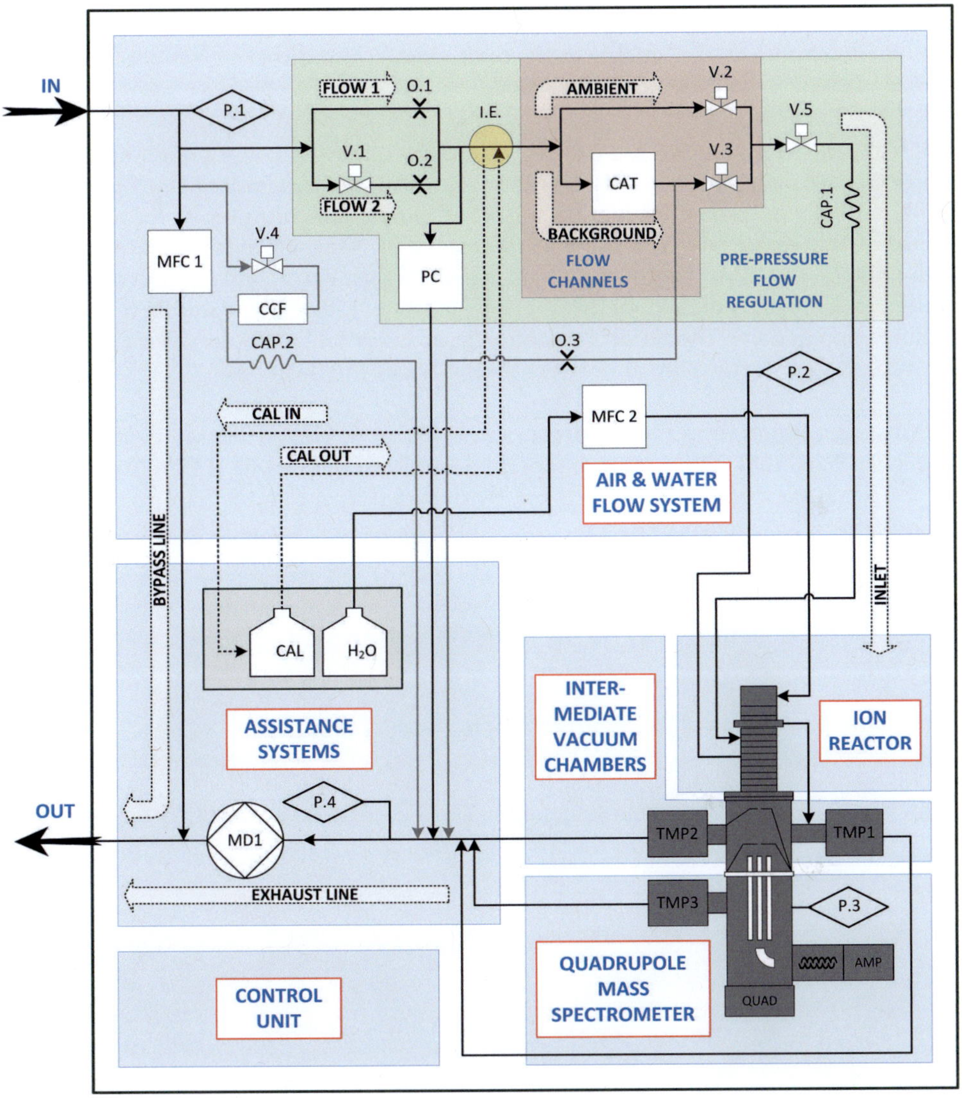

Figure F 3.3: *ULW-PTR-MS system overview (w/o cabling/electronics) showing the functional units (as denoted in the graph; cf. paragraph 'Units' in section 3.1 on page 40). Used abbr.:* ***AMP*** *amplification,* ***CAL*** *calibration unit,* ***CAP.x*** *capillary,* ***CAT*** *catalyst,* ***CCF*** *char coal filter,* $\boldsymbol{H_2O}$ *water reservoir,* ***I.E.*** *isotopic enrichment (optionally applicable),* ***MD1*** *diaphragm pump,* ***MFC*** *mass flow controller,* ***O.x*** *orifice,* ***P.x*** *pressure sensor,* ***PC*** *pressure controller,* ***QUAD*** *quadrupole,* $\boldsymbol{TMP_{\#x}}$ *turbo molecular pump,* ***V.x*** *valve.*

and allows optional upgrading in the future. Moreover, fast adaptation to mission-specific demands is possible due to the open design and direct accessibility of hardware components and developed software.

The V25 system is inserted to be either manually operated using the integrated operator control panel with display on the front side of the instrument (see figure F 3.4, p. 42), or fully automated when deployed at remote and unmanned locations (e.g. inside of the cargo bay of the CARIBIC aircraft). Measurement data, parameter files and the executed program copies are stored on extractable flash cards (data memory: 256 MB; typical data rates $< 1\,MB \cdot hour^{-1}$), that are readable directly via tethered access using the serial connection to an external DOSBox[14] / LabVIEW[15] environment, or by removing and external readout of the flash disc.

A flow chart illustrating typical process operations of ULW-PTR-MS is attached to appendix B (p. 163). The processing of measurement data using different strategies for the evaluation of single peaks is briefly outlined in appendix C (p. 165).

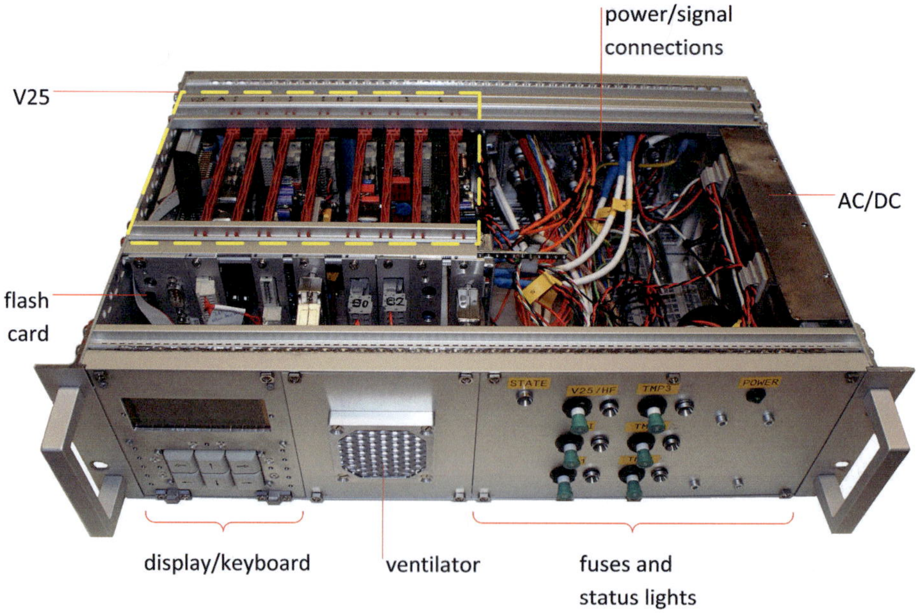

Figure F 3.4: *The Control Unit in plan view after opening of the top covering. Operating elements on the front side of the 19″ subrack and other components are labeled.*

14 DOSBox → `http://www.dosbox.com/`

15 National Instruments LabVIEW → `http://www.ni.com/labview/`

MODULE	IDENTIFIER	PURPOSE/control of	# of modules
• **analog voltage input**	AIN8X	temperature sensors (8x) (table T 3.2, p. 54)	2
• **analog voltage output**	DCDC8X	heatings (table T 3.2, p. 54); ventilation; pump control MD1; valves V.1 - V.5	2
• **interface to external controllers/sensors**	MFC6X	$MFC_{\#1}$; $MFC_{\#2}$; PC; pressure sensors: P.1 (INLET), P.2 (DT), P.3 (QUAD)	1
• **(high-)voltage supplies**	HV6X/HV2X	ion reactor (IS, ISDR, DT); NC; QMS; SEM	5
• **mass spectrometer control**	MS	triggering of the high-frequency generator; AMP	double/2
• **RS 485** (serial bus port)	V25SER	control of $TMP_{\#1}$, $TMP_{\#2}$, $TMP_{\#3}$	(extra)
• **network**	V25XPORT	data line (here: to HALO)	(extra)
• **host**	COM	(optional) tethered connection to DOSBox/LabVIEW	(extra)

Table T 3.1: *V25 mounting with register of the used modules (module specifications in MPIC [2012]).*

Powering The powering is essentially adapted to the power conditions in the HALO aircraft, which are governed by the aircraft power supply with its 3-phase AC voltage of $110\,V$ at a frequency of $400\,Hz$ [DLR, 2009]. The instrument's input voltage is transformed by internal AC/DC ($U_{system} = 115 - 230\,VAC$ (line frequency $47 - 440\,Hz) \xrightarrow{\text{AC/DC}} 24\,V$) and DC/DC voltage converters ($24\,V \xrightarrow{\text{DC/DC}} -24\,V$; $24\,V \xrightarrow{\text{DC/DC}} 5\,V$) and fuse-protected (main $10\,A$; individual $1 - 7.5\,A$). The earth star point connects both the input earth (via the input filters) and the instrument frame as base reference, which avoids ground loops and with that leakage currents causing interference fields.

3.1.2 Air & Water Flow System

The Air & Water Flow System handles the sample air and the H_2O vapor. Figure F 3.3 (p. 41) shows the particularized setup in form of the system diagram and a corresponding top view picture of the flow control is presented in figure F 3.5 (p. 44). The unit is being installed into a 19″ subrack.

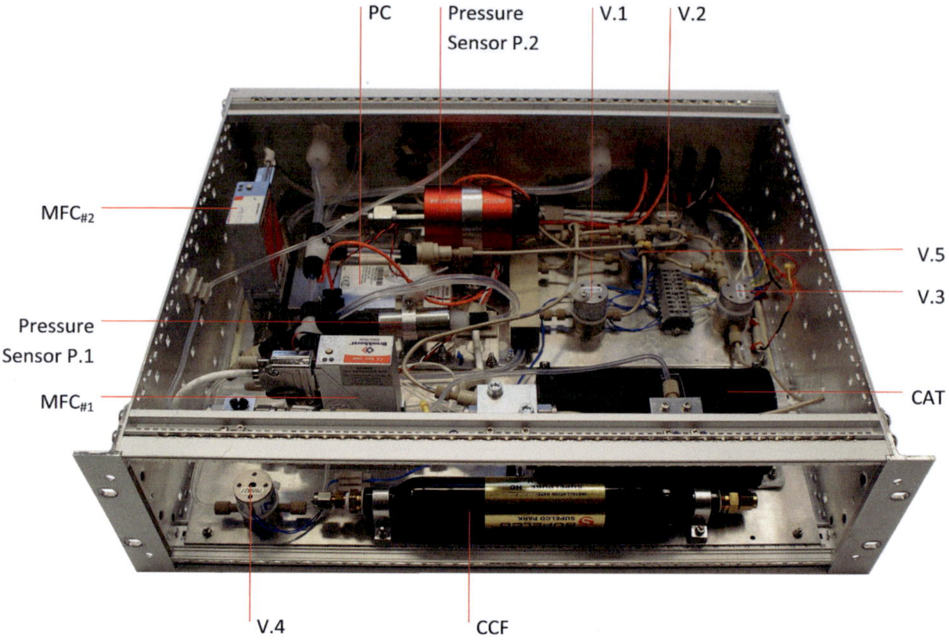

Figure F 3.5: *The Air & Water Flow System in plan view after opening of the top and front covering. Labels refer to the system overview in figure F 3.3 (p. 41).*

System requirements and setup For the premise of atmospheric measurements under strongly varying external pressures during aircraft missions, that range from $P_{ambient} \simeq 1'013\,hPa$ at sea level to $P_{ambient} \simeq 120\,hPa$ (ICAO[16] standard atmosphere) at the HALO flight ceiling altitude $H_{flightceiling} = 15'000\,m$, a pressure regulation prior to the mass spectrometer is installed. The entire inlet system is thermally controlled in order to avoid memory effects by absorbing and desorbing of constituents on the inlet line walls, and condensation of H_2O vapors. Likewise, with this thermal control all temperature sensitive devices (like P and F controllers) are kept on stable levels (see table T 3.2, p. 54). In order to reduce perturbing memory effects, mostly PEEK (Polyether Ether Ketone) or PFA (PerFluoroAlkoxy) is used for tubings, connectors and housings. These materials have been found to be as inert as possible for the reported VOCs due to their low sorption behaviors [Dallas & Carr, 1991; Krüger *et al.*, 2014]).

Pressure regulation The sample air is transported to the instrument through a $^1/_4$″ PFA tube from the outside. Pressure sensor P.1 monitors the ambient pressures $P_{ambient}$. An upstream bypass flow $F_{bypass} = 1\,ln\cdot min^{-1}$ controlled by mass

16 International Civil Aviation Organization

flow controller $MFC_{\#1}$, from which the instrument aspirates $\sim 100\,mln\cdot min^{-1}$ (at ground pressures), guarantees short inlet time delays ($\leq 1\,s$) and comparable response times. The main flow downstream of $MFC_{\#1}$ is guided to the instrument exhaust line, which is then driven to the main HALO exhaust line alongside the instruments (when deployed on board). Likewise, the ram pressure in connection with airborne measurements essentially drives the bypass line. The extracted air to be measured is branched off between P.1 and $MFC_{\#1}$, where the following feeding line is reduced to $^1/_8$" PFA and PEEK tubings.

When turning off the instrument or in case of fault (after shutdown of the pumps) valve V.4 opens to enable a venting of the high vacuum chambers with ambient air via the downstream side of then decelerating turbo molecular vacuum pumps ($TMP_{\#1,\#2,\#3}$). An activated charcoal filter (CCF) for additional cleansing from VOCs [Metts & Batterman, 2006] can be optionally interconnected. The capillary CAP.2 (material: PEEK; ID $\sim 213\,\mu m$; OD 1/16"; $L_{CAP} \simeq 0.9\,cm$) reduces the flow in this venting line.

The pressure in the DT is controlled by pressure controller PC in combination with critical PEEK orifices O.1 ($ID_{O.1} = 120\,\mu m$) & O.2 ($ID_{O.2} = 180\,\mu m$) and a capillary (CAP.1) (material: PEEK; ID $\sim 250\,\mu m$; OD 1/16"; $L_{CAP} \simeq 1\,cm$) as flow restrictor. The system is attuned to maintain the pre-pressure $P_{pre} \sim 70\,hPa$ at master valve V.5 in order to have an inflow of $F_{inflow} \sim 25\,mln\cdot min^{-1}$ into the DT. The drift tube pressure then results to $P_{drift} = 2.3\,hPa$ (measured by P.2). The permanently open orifice O.1 allows reasonable rates of flow (FLOW 1) at lower altitudes or higher pressure, respectively. Under the premise that the inflow and pressure in the DT is retained, the pressure controller PC leaves its manageable ranges of $0 - 150\,mln\cdot min^{-1}$ at low ambient pressures (high flight altitudes). Therefore, the second orifice O.2 is connected by additional opening of V.1 in order to increase the inflow (FLOW 1 & FLOW 2). The appropriate apertures of O.1 and O.2 are individually selected to allow measurements across the entire pressure range from ground pressures down to $120\,hPa$ in consideration of the limited pumping efficiency of the backing pump MD1 (maximum pumping speed $\sim 170\,mln\cdot min^{-1}$; fore pressure limit $10\,hPa$). The switch over to a combined flow via both channels is configured to happen at ambient pressures of $P_{ambient} = 400\,hPa$, which corresponds to an altitude of $\sim 7.2\,km$ (ICAO standard atmosphere). For an adaptation to other pressure conditions, orifices and capillary are to be exchanged.

After the T-junction to controller PC and before being injected into the flow channel section, the air can be optionally enriched with isotopically labeled compounds for on-line calibration (see section 3.1.6, p. 52). Likewise, a feeding of admixed sample gas standards using gas cylinders or permeation ovens can also be performed at this connecting line during external system calibrations (see section 4.1, p. 57).

Catalytic cleansing of sample air Between the pressure regulators and master valve V.5, the air flows through either of the two channels AMBIENT and BACKGROUND controlled by reciprocal opening of valves V.2 and V.3. The AMBIENT channel is used for standard measurements with the detection of all existing VOCs in the sample air. When using the BACKGROUND channel only system-internal background signals are detected, as the injected ambient air is additionally guided through the heated platinum catalyst (CAT) for VOC scrubbing. This process of catalytic oxidation happens at temperatures of $T_{catalyst} \sim 350\,°C$ and was reported to have an efficiency of $\gg 95\,\%$ for VOCs and other gas-phase organics [Kosusko & Nunez, 1990]. A typical measurement cycle comprises $3\,min$ background signal measurements every $30\,min$. With that, the concentration of VOCs in the sample air is then derived by direct comparison of both channels in form of a subtraction of background signals from the ambient signals. The background signals are linearly interpolated between adjacent background measurements. Low background rates indicate a clean inlet system. In order to avoid memory effects in the catalyst channel, air is permanently flushed through an open small orifice O.3 ($ID_{O.3} = 100\,\mu m$) towards the general exhaust line.

Supply of H_2O vapor to the IS The H_2O vapor originating from the headspace of the liquid water reservoir is controlled by $MFC_{\#2}$ (low-Δp flow controller; $15\,mln\cdot min^{-1}$) using typical flows of $F_{H_2O} = 7.5\,sccm$ (see section 3.1.6, p. 51). The connecting line is a $^1/_4$″ PFA tube, which has an intermediate drip chamber consisting of a $^3/_8$″ PFA tube in order to suppress formed drops in the line from climbing to the water-sensitive MFC, which can emerge due to condensation and capillary effects in the tubing. All lines are additionally heated for the same reason.

3.1.3 Ion Reactor

The ion reactor is the central unit facilitating the ionic reactions of H_3O^+ formation and proton transfer with VOCs (see chapter 2.2, p. 15). It comprises the ion source (IS) and the drift tube (DT), which are connected through the intermediate ion source drift region (ISDR). A schematic diagram of the ion reactor mounted on top of the quadrupole mass spectrometer is presented in figure F 2.1 (p. 14).

Ion source (IS) The custom-built IS, which generates H_3O^+ ions (see section 2.2.1, p. 15), comprises two connected anodes and the cathode positioned in between. It is manufactured from stainless steel with intermediate insulating PEEK-rings. In the upper anode an inlet is inserted to inject the incoming H_2O vapor. The bottom anode has a critical orifice ($ID = 0.6\,mm$) merging into a nozzle, that extracts the ions into the ISDR. The electrodes are supplied by high-voltage enabling an electrical breakdown between the different potentials at anodes (typical voltage: $U_{anodes} \sim 1\,kV$) and cathode (typical potential difference $U_{anodes} - U_{cathode}$: start $\sim 600\,V$; working $\sim 430\,V$), which leads to the formation of an electrical discharge that generates a plasma. This hollow cathode discharge is stabilized by pre-set electrical currents of typically $I_{IS} = 3 - 5\,mA$ and also depends on the geometry of the chamber itself [Hartungen, 2005; Howorka *et al.*, 1973; Slevin & Harrison, 1975].

The here used most recent home-built version allows rates of $i(H_3O^+) \lesssim 1.7 \times 10^7\,cps$ (observed at the detector after DT and mass filter). Commercially available IS from IONICON in Innsbruck/Austria (most recent version from 2009) provide reagent ion rates of $i(H_3O^+) \lesssim 4 \times 10^7\,cps$ at the detectors of the standard instrument, that has a more powerful HV pump and larger DT apertures [Jordan *et al.*, 2009a].

The IS is typically tuned to ion yields of NO^+ (m/z 30) $\leq 2\,\%$, O_2^+ (m/z 32) $\leq 2\,\%$ and $H_3O^+(H_2O)$ (m/z 37) $\sim 10\,\%$ of the free H_3O^+ (m/z 19) ions in the DT. Inflows of H_2O vapor (using $MFC_{\#2}$), applied electrode voltages and corresponding pumping rates at the downstream ISDR are concerted for that (see next paragraph).

Ion source drift region (ISDR) When entering the ISDR, the ion beam still contains ions that are not fully recombined to H_3O^+ (see hydronium reaction chain presented in table T 2.1, p. 15). These ions can cool and relax to H_3O^+ in the ISDR. The kinetic energies of the ions are controlled by the applied potentials of typically $\Delta U_{ISDR} = 120 \pm 20\,V$ across this chamber determining acceleration and with that sojourn times of H_3O^+ ions. This affects the broadening of the H_3O^+ ion beam and thus the number of primary ions entering the DT. In addition, also the amount of contaminant ions in consequence of reactions with back-streaming sample air in the IS are affected. Moreover, as not ionized H_2O molecules can further collide with other ions/molecules forming clusters or fragmenting them, remaining H_2O molecules originating from the IS are pumped away from the ISDR before being injected into the DT. In order to control the flow rate via the pumping line between ISDR and $TMP_{\#1}$ (see figure F 2.1, p. 14), a manually adjustable hose clamp is inserted.

By improving the ability of water, O_2^+ and NO^+ removal without disproportionate manipulation of the ion beam and prevailing pressures P_{IS} and P_{ISDR}, the quality of PTR measurements by means of increased total sensitivities are expected to raise. A redesign of this compartment under special consideration of the pumping concept is therefore recommended.

Drift Tube (DT) The DT follows after the ISDR and is separated by an aperture (critical orifice, $ID = 0.5\,mm$) extracting the primary ions into the reaction chamber, where the PTR between primary H_3O^+ ions and injected VOCs happens. The DT (length $L_{drift} = 8.8\,cm$) is built up consisting of an alternating sequence of 10 stainless steel rings, that are insulated to each other using PEEK spacers with Viton O-ring gaskets. Into the uppermost ring sample air is fed from the inlet system. The DT pressure is monitored by pressure sensor P.2 and is attuned to be $P_{drift} = 2.2\,hPa$ (optimal pressure range: $P_{drift} = 2.0 - 2.4\,hPa$ [de Gouw & Warneke, 2007]). P_{drift}

is regulated using the described upstream pre-pressure regulation in combination with the pumping from the downstream intermediate vacuum chambers (see complete overview in figure F 2.1, p. 14). Due to possible influences onto the PTR reactions, the DT is thermally controlled to temperatures of $40 - 60\,°C$.

An interconnected chain of 9 single consecutive resistors with resistance $R = 150\,k\Omega$ attached along the steel electrodes form a voltage cascade over the drift region from the very top to the bottom. With applying the higher voltage to the uppermost electrode and the lower voltage to the bottom plate, the linearly decreasing potentials establish a homogeneous electric field inside the DT with $\Delta\Phi_{first \mapsto last}$ typically $500 - 700\,V$. Besides influencing the ions' kinetic energies and with that the cluster-ion formation (as discussed in section 2.2.3, p. 20), the E-field controls the residence times ($\equiv$ reaction times) of the ions in the DT ($\sim 100\,\mu s$). The ions are transported through the DT by dragging their positive charges along the field gradient towards the lower aperture ($ID = 1.2\,mm$) at the downstream end. Typical operating values for standard measurements with the ULW-PTR-MS lead to an electric field strength E (in $[V \cdot cm^{-1}]$) of:

$$E = \frac{\Delta\Phi_{first \mapsto last}}{L_{drift}} \tag{E 3.1}$$

$$= 69.5\,V \cdot cm^{-1} \tag{E 3.2}$$

with:

$\Delta\Phi_{first \mapsto last}$:	potential difference	$\Delta\Phi_{first \mapsto last} = 612\,V$ (measured)
L_{drift}:	drift tube length	$L_{drift} = 8.8\,cm$ (measured)

The gas density in number of molecules N per volume V (here: $[cm^3]$) is calculated using the ideal gas law ($PV = nRT = Nk_BT$ with number of moles $n = N/N_A$, molar volume $V_{M0} = RT_0/P_0$ and gas constant R[17] $= k_B$[18]$\cdot N_A$[19]):

$$N = \frac{P_{drift}}{T_{drift}} \cdot \frac{T_0}{P_0} \cdot \frac{N_A}{V_{M0}} \tag{E 3.3}$$

$$= 5.091 \times 10^{16}\,cm^{-3} \tag{E 3.4}$$

with:

P_{drift}:	DT pressure	$P_{drift} = 2.2\,hPa$ (measured)
T_{drift}:	DT temperature	$T_{drift} = 313\,K$ (measured)
P_0:	reference pressure	$P_0 = 1013.25\,hPa$
T_0:	reference temperature	$T_0 = 273.15\,K$
N_A:	Avogadro constant	$N_A = 6.022 \times 10^{23}\,mol^{-1}$
V_{M0}:	molar gas volume	$V_{M0} = 22.414 \times 10^3\,cm^3 \cdot mol^{-1}$

17 Molar gas constant $R = 8.3144621\,J \cdot mol^{-1} \cdot K^{-1}$(NIST)
18 Boltzmann constant $k_B = 1.3806488(13) \times 10^{-23}\,J \cdot K^{-1}$(NIST)
19 Avogadro constant $N_A = 6.02214129(27) \times 10^{23}\,mol^{-1}$(NIST)

The result $N = 5.091 \times 10^{16}\, cm^{-3}$ is then used to derive the ratio ${}^{E}/_{N}$ (for H_3O^+-ions):

$$\frac{E}{N} = \frac{69.5\, V \cdot cm^{-1}}{5.091 \times 10^{16}\, cm^{-3}} \qquad \text{(E 3.5)}$$
$$= 1.365 \times 10^{-15}\, V \cdot cm^2 \qquad \text{(E 3.6)}$$
$$= 136.5\, Td \qquad (Td = 10^{-17}\, V \cdot cm^2) \qquad \text{(E 3.7)}$$

The calculated value ${}^{E}/_{N} \sim 137\, Td$ is in the upper range of typical preferences for PTR-MS instruments (${}^{E}/_{N} = 100 - 140\, Td$ [Blake *et al.*, 2009]) and is a compromise between low clustering rates and maximum reaction times at the expense of possible ion fragmentations and with that slight reduction of the instrumental sensitivity (further discussed in section 2.2.3, p. 20).

Off-axis adjustment of the ion reactor Photons disturb the ion counting in the QMS when stimulating electrons to release from the detector electrodes (photo-electric effect), which would increase background signals and with that cause a reduction of the sensitivity. For that the entire ion reactor is slightly bended so that photons originating from the glow discharge in the IS are blocked and cannot straightly traverse the vacuum chambers into the sensitive secondary electron multiplier (SEM) detector.

3.1.4 Intermediate Vacuum Chambers

Setup The intermediate vacuum chambers (IVC) follow the lower aperture of the DT and comprise two interstages prior to the high vacuum chamber. Each of the two are connected to a turbo molecular pump (see figure F 2.1, p. 14). The chambers are separated among each other by another aperture. This nose cone plate (NC) ($ID = 1.3\, mm$) is supplied by a voltage of $U_{NC} = 57\, V$. As this voltage is typically below the two-part DT bottom plate potentials of $U_{bottom\uparrow} = 68\, V$ and $U_{bottom\downarrow} = 73\, V$, respectively, the cations are dragged out of the DT.

Enhancement of the mean free path towards the QMS Due to pumping with turbo molecular pumps $TMP_{\#1,\#2}$ at these chambers, the pressures reduce from $P_{drift} = 2.2\, hPa$ at the head to $P_{quadrupole} = 3 \times 10^{-5}\, hPa$ at the bottom of this pressure gradient. This leads to an increase of the corresponding mean free path λ:

$$\lambda = \frac{RT}{\sqrt{2}\pi d^2 N_A P} \qquad \text{(E 3.8)}$$

with:

R:	universal gas constant	$R = k_B \cdot N_A$ (→ equation E 3.3)
T:	temperature	$T = 300\, K$
d:	diameter of air molecules (ideal gas)	$d_{Loschmidt} = 3.61 \times 10^{-10}\, m\ (\AA)$
P:	pressure	1.: $P_{drift} = 2.2\, hPa$ 2.: $P_{quadrupole} = 3 \times 10^{-5}\, hPa$

from $\lambda_{drift} \cong 32\,\mu m$ to $\lambda_{quadrupole} = 2.4\,m$. With that the previously ionized trace gas composition is mostly conserved for the following detection in the high-vacuum quadrupole mass spectrometer as further collisions with remaining air molecules are unlikely to happen.

Another degree of freedom: Ion acceleration/deceleration in the IVC Depending on the actual ion velocities ν_{ion} (see equation E 2.4, p. 20) when entering the intermediate sections, the ions typically undergo acceleration according potentials applied to the DT bottom plate and the extraction lenses NC plate and following guiding lenses (depicted as ion optics in figure F 2.1, p. 14). Nevertheless, it has been found that a smooth ion deceleration between the two-part bottom plates (typical settings: $U_{bottom\uparrow} = 68\,V$; $\Delta U_{bottom(\uparrow-\downarrow)} = -5\,V$) directly below the DT increases the ion count rates of some compounds at the detector in a nonspecific manner. It is assumed that due to the deceleration of injected ions the ion's residence time slightly extends, which then enables further reactions between ions and remaining air molecules. The latter are then pumped away and the ions undergo once more acceleration towards the injecting ion optics above the quadrupole.

However, as this effect is not quantitatively studied by now, this implemented additional degree of freedom for fine-tuning of the instrument needs further investigation.

3.1.5 High Vacuum Quadrupole Mass Spectrometer

High-vacuum chamber In the high vacuum (HV) the mass filtering by the quadrupole mass spectrometer and the detection by the secondary electron multiplier (SEM) occur. The used standard devices have been adapted and installed to the custom-built vacuum chamber manufactured from aluminum (alloy 5020) by precision machining, welding and honing [Brito, 2011]. The HV chamber is pumped by $TMP_{\#3}$ to typical pressures of $P_{quadrupole} = 2 - 3 \times 10^{-5}\,hPa$ during measurements and $P_{quadrupole} < 9 \times 10^{-7}\,hPa$ without gas ballast. The HV pressures are measured by pressure sensor P.3.

Quadrupole mass filter The ion injection into the HV system consists of the two-lens ion optics in front of a grounded entry aperture, which are immediately followed by the quadrupole array consisting of four molybdenum rods with diameters $r_{rods} = 8\,mm$ and lengths $L_{rods} = 200\,mm$ (see schematic setup in figure F 2.2, p. 27). With that setup the ions are collimated to a virtually parallel beam before entering the quadrupole field (further discussed in section 2.3, p. 26). The resolution and mass range of quadrupole mass spectrometers strongly dependent on the geometric parameters of the rod system, the radio frequency supply ($1.5 - 2350\,Vp$, $\omega_{HF} = 2.25\,MHz$) and the ion injection energies ($V_z \approx 5\,eV$ [Dawson, 1976]).

An estimated maximum instrumental resolution ΔM (being the FWHM of the peak at mass M in $[amu]$) can be calculated according:

$$\Delta M = \frac{4 \times 10^9 V_z}{\omega_{HF} \cdot L_{rods}} \qquad \text{(E 3.9)}$$

with:

ΔM:	width of peak at mass M (e.g. FWHM)	$[amu]$
V_z:	ion injection energy	$V_z \approx 5\,eV$
ω_{HF}:	frequency of HF generator	$\omega_{HF} = 2.25\,MHz$
L_{rods}:	length of rods	$L_{rods} = 200\,mm$

The calculated minimum obtainable peak width is $\Delta M = 0.099\,amu$. Nevertheless, the expected resolution for the used combination of generator and rod array is lower due to mechanical imperfections, like e.g. slight misalignments of the rod system, and should be $0.3\,amu$. In practice, the resolution using a standard quadrupole is often set to be close to unity, i.e. $\sim 1\,amu$.

$90°$ **ion deflection** After the quadrupole array the remaining target ions, that have passed the quadrupole field on stable trajectories, get deflected by $90°$ via a combination of two deflection electrodes towards the detector (see figure F 2.1, p. 14). Fast or excited neutrals and still remaining photons keep their initial flight paths towards the dead end of the chamber and with that do not induce additional noise at the detector.

Detector The used secondary electron multiplier (SEM) consists of 17 dynodes (material: Cu-Be) and is supplied by high-voltages of up to $U_{detector} \leq 3.5\,kV$, which needs to be regularly adapted according to downgrading effects of the electron amplification in consequence of aging and coating. The corresponding gain is $> 10^8$ at $U_{detector} = 3.5\,kV$ [Pfeiffer-Vacuum, 2007] and typical discrimination thresholds of $0.3\,V$ for registering of incoming ions are used. The SEM has a pulse width of $\sim 10\,ns$ and a double pulse temporal resolution of $\leq 20\,ns$.

3.1.6 Assistance Systems

Diaphragm pump MD1 The backing pump MD1 is connected to the back-pressure side of the inlet system via pressure controller PC and to the turbo molecular pumps $TMP_{\#1,\#2,\#3}$ (see figure F 3.3, p. 41). The upstream pressure of the MD1 is monitored with pressure sonsor P.4. During solely use of this pump, chamber pressures of $P_{drift} < 2\,hPa$ are obtained.

Radio frequency generator The DC and AC (frequency $\omega_{HF} = 2.25MHz$) voltages for controlling the QMS are provided by a high frequency generator and are coupled onto the rods of the quadrupole mass spectrometer (described in detail in section 2.3, p. 26).

H_2O tank/calibration unit A combined water reservoir and embedded permeation unit for on-line calibration is inserted, which is discussed in detail in Brito & Zahn [2011]. It is built up of welded stainless steel and takes hold of up to $500\,ml$ of distilled water. The unit is thermally decoupled from the exterior by covering with vacuum pads, which enables the very slow adaptation of the water temperature in the inside to ambient temperatures. With that thermal fluctuations in the exterior do not directly affect the reservoir temperature and provide mostly stable conditions in the inside. The primary task of this unit is the supply of H_2O vapor to the IS serving as feedstock for the H_3O^+ formation (described in section 3.1.2, p. 43). In addition, its simultaneous use as unheated VOC permeation source using isotopically labeled compounds as internal calibration standards is particularly enabled due to the temperature stability inside of the containment. This feature is essential, since permeation rates for VOCs across the inserted thin teflon membrane strongly depend on temperature. The used compounds (by now: acetone-d6) are filled to embedded but separately closed pipes. When interconnected to the inlet line, the air to be measured is transported through an exchange area including the permeation membrane for the specific enrichment (see figure F 3.3, p. 35). The used deuterated compounds do not naturally occur in ambient air but appear on corresponding altered m/z in the PTR mass spectra, like e.g. acetone-d6 on m/z 65 instead of m/z 59. Continuous monitoring of affected m/z enables a permanent specific on-line calibration of the instrument also during mission operations and efficiently complements the standard pre- and post-measurement calibrations using external gas cylinders (see section 4.1, p. 57).

3.2 Aspects of the electrical and mechanical engineering and temperature management

In accordance with the requirements of the aeronautical certification for the deployment aboard the HALO aircraft, the ULW-PTR-MS has been set up in consideration of general key aspects concerning the electrical and mechanical engineering:

1. **Electromagnetic compatibility (EMC):** In order to prevent crosstalk with aircraft systems (like e.g. a disturbance of the radio system) or other scientific equipment, electromagnetic interferences (EMI) have to be suppressed. The appropriate frequency dependent thresholds are regulated by the international standard RTCA/DO-160G[20] (see also section 5.2.1, p. 86).

   - **Input/Output filtering:** The power connection between the HALO on-board power supply and the instruments' AC/DC converters (capable for $U_{system} = 110 - 250\,V$) is separated using noise suppression filtering with an operating AC voltage of $250\,V$ and a frequency of $0 - 400\,Hz$.
   - **Shielding:** Radiating and sensitive components are installed in boxes with EMI sealings. In addition, perforated aluminum plates as protection skirts are mounted around sensitive areas (e.g. backside of the instrument pointing to the wiring in the aircraft frame). Potential antenna effects are suppressed using drilled/shielded cabling.

2. **Mechanical strength:** The mechanical engineering is strongly influenced by certification guidelines.

   - **Mechanical loads:** Load-bearing structures (e.g. holding frames, mounting brackets) are made of light high-strength aluminum alloy, that is approved for airborne applications (material identification number: 3.4364).
   - **Shock damping:** In order to prevent from shocks and vibrations, particularly sensitive equipment, like e.g. the high-frequency generator, is mounted on spiral springs (wire rope isolators) for appropriate shock absorption.

3. **Fire protection/Temperature management:** Fire protection requirements as well as the perpetuation of stable operations of the system necessitate the surveillance of temperatures. Therefore NTC[21] resistors and a PT100 sensor[22] inside the platinum scrubber are inserted for monitoring. Some temperatures are actively controlled by heating/ventilating according the set-point temperatures listed in table T 3.2 (p. 54) and supplemented using irreversible thermal fuses for safety shutdown.

[20] Environmental Conditions and Test Procedures for Airborne Equipment; published by Radio Technical Commission for Aeronautics: `http://www.rtca.org/`

[21] Negative Temperature Coefficient Thermistor

[22] Platinum resistance temperature detector Pt100 ($R_0 = 100\,\Omega$)

	AREA	TEMPERATURE (regulated to)	TYPE
HEATING	**Ion reactor**	$T_{reactor} \simeq 40 - 60\,°C$	heating wire; insulation cover
	Platinum scrubber	$T_{scrubber} \simeq 350\,°C$	heating cartridge; insulating foam
	H_2O vapor line	$T_{H_2Ovaporline} \simeq 30\,°C$ ($T_{H_2Ovaporline} > T_{ambient}$)	heating wire
	Ambient air inlet line	$T_{inlet} \simeq 40\,°C$ ($T_{inlet} > T_{ambient}$)	heating wire
	Flow Box (complete)	$T_{FlowBox} \simeq 40\,°C$ ($T_{FlowBox} > T_{ambient}$)	heating plate; insulation cover
	HALO - TGI (Trace Gas Inlet)	(mission-specific/ anti-freeze)	(inlet heater powered via the instrument)
COOLING	**Control Box** (complete)	(anti-overheating)	fan
	AC/DC converters	(anti-overheating)	radiator fins
	Pumps (MD1, $TMP_{\#1,\#2,\#3}$)	(anti-overheating)	fans

Table T 3.2: *Surveillance and regulation of instrument temperatures with specification of temperature ranges and appropriate heating and cooling devices.*

3.3 Fields for improvements and add-ons

Although the ULW-PTR-MS and its sister instrument have already been applied operationally, the requirements for highly sensitive measurements demand further improvements of the system/process. Some areas with potentials for improvements (as already above-mentioned) are briefly summarized in table T 3.3 (p. 55).

OBJECTIVE	AREA	ACTION
increase of primary ion yield $i(H_3O^+)$ ⇒ enhancement of sensitivity	**IS, ISDR**	refinement of IS, ISDR geometry and settings
reduction of O_2^+, NO^+ molecules ⇒ purification/"cleaning" of H_3O^+ primary ion beam	**ISDR**	redesign of the ISDR
reduction of remaining H_2O molecules ⇒ reduction of clustering/fragmentation	**ISDR**	redesign of the ISDR
improvement of transmission efficiency of ions exiting the DT ⇒ general increase of ion rates (→ sensitivity)	**DT**	implementation of further ion focusing electrodes at the bottom end of the DT
improvement of ion transmission efficiency in the IVC ⇒ increase of ion count rates (→ sensitivity)	**IVC**	investigation of effects of ion acceleration/deceleration between DT and QMS; fine-tuning
reduction of background noise ⇒ improvements of accuracy and LOD	**IS, IVC**	fine-tuning of ion guiding, pressures and QMS settings

Table T 3.3: *Some areas for instrumental improvements portraying actions, that are recommended to be pursued in the future.*

Starting points for add-ons comprise:

- **Implementation of an ion funnel to the DT:** With the implementation of an ion funnel for specific ion guiding and focusing in the DT, the primary ion count rates are expected to clearly increase. This has been demonstrated using Monte Carlo[23] simulations of ion trajectories across the reaction zone and laboratory studies [Brito, 2011; Kelly *et al.*, 2010]. Nevertheless, such modifications, although being taken into consideration for the future, have to be considered wisely as this

[23] Numerical calculations by means of statistical random sampling

challenges the use of the instrument. Compatible RF technology enabling an acceptable focusing especially of smaller m/z, which is necessary as long as H_3O^+ ions are used for normalization purposes, is expected to be difficult to make available and embed within the used V25 system.

- **Setup of a compatible laboratory calibration facility:** In order to complement and extend the existing calibration capabilities (see also section 4.1, p. 57), a planned laboratory calibration facility comprising devises for appropriate dilution and humidification of the diverse applicable calibration gases is to be set up. Especially with regard to the intended investigations of compounds anticipated to be observed in very low concentrations in higher altitudes, a regular use of such facility is recommended to obtain comparable data sets.

Chapter 4

Instrumental Characterization

The ULW-PTR-MS was deployed side by side to high-sensitivity VOC instruments for weeks during the mission UBWOS 2012. Used reference systems were the PTR-MS [de Gouw & Warneke, 2007] and the gas chromatography mass spectrometer (GC-MS) [Gilman *et al.*, 2009; Goldan, 2004] from the National Oceanic And Atmospheric Administration (NOAA) in Boulder, Colorado/USA. While being operated in parallel comparative measurements investigating the system properties of ULW-PTR-MS have been performed accompanying to the scheduled measurements. Both PTR instruments measured a largely consistent set of compounds (see table T 4.2, p. 64) and the GC-MS measured, among others, speciated C_2-C_{12} alkanes, C_5-C_8 cycloalkanes, C_6-C_9 aromatics and methanol.

The ULW-PTR-MS was in an interim version during the campaign with an instrumental sensitivity of $100 - 200\,{}^{cps}/_{ppbV}$. In the context of the system upgrade to the version used aboard the HALO aircraft, changes have been made comprising mainly advancements of the mechanical and electrical setup due to requirements of the aeronautical certification. With the installation of a smaller volume IS (see paragraph 'Ion source (IS) efficiency' in section 3.1.3, p. 46) and further fine-tuning of DT and quadrupole settings, the primary ion rate has been increased by a factor of ~ 3 since then and the instrumental sensitivity increased to $300 - 600\,{}^{cps}/_{ppbV}$ on average. Despite everything, it is worth mentioning that derived calibration parameterization is still applicable as it primarily depends on geometric prerequisites and other factors influencing the PTR (like pressures, DT potentials), that have not been changed since then.

4.1 System calibration

The applied calibration techniques comprise the measurement of differently diluted calibration gas mixtures originating from pressurized cylinders, that are admixed to either ambient or synthetic air flows using mass flow controllers for precise dosing. Likewise, permeation tubes as gas sources for compounds are used, that are not stable in gas cylinders/mixtures due to catalytic effects on walls, polymerization or reactions with other ingredients. Compounds with water-sensitive calibration factors have been handled separately in consideration of appropriate moisturization ranges. The derived calibration factors S or system sensitivities ε_R, respectively, which are the slopes of mixing ratios vs. ion count rates, are declared in units of $[^{ncps}/_{ppbV}]$, meaning the to primary signals normalized counts per second per concentration given in parts per billion (10^{-9}) by volume (see also section 2.2.3, p. 20).

During UBWOS 2012 the gauged NOAA gas standard AAL071532 as primary calibration gas source has been used. In addition, the Mobile Organic Carbon Calibration System (MOCCS) [Veres *et al.*, 2010] has been utilized to calibrate to sample gases originating from permeation tubes, like formaldehyde (HCHO). MOCCS combines the production of gas-phase VOC standards with quantitative conversion of total organic carbon to CO_2 in a palladium catalyst at $350\,°C$ under the presence of oxygen, and the subsequent CO_2 measurement using a LI-COR analyzer (LI-6252) (NDIR gas analyzer using a $4.26\,\mu m$ absorption band for CO_2). Flows have been controlled using MFCs (Bronkhorst EL Flow) and were individually measured volumetrically by a primary gas flow calibrator (Bios DryCal Definer 220).

Note: Multiple assignments of m/z values in PTR-MS mass spectra The calibration standards typically consist of selections of stable/nonreactive compounds. Ambient air on the other hand consists of varying compositions of diverse organic compounds naturally comprising isomers, isobars (and isotopologues). In PTR-MS technique the additional difficulty of appearing fragments of compounds occurs, like e.g. on m/z 79, where higher aromatics start fragmenting when working with higher E/N ratios [de Gouw & Warneke, 2007; Yuan *et al.*, 2014]. All these effects lead to a possible multiple assignment of compounds to identical m/z in a mass spectrum, like e.g. m/z 59, where the isomers acetone (C_3H_6O; PA(acetone) = $812.0\,kJ\cdot mol^{-1}$) and propanal (C_3H_6O; PA(propanal) = $786.0\,kJ\cdot mol^{-1}$) are detected (see also table T 2.2, p. 18). In addition, other interferences between compounds depending on their relative abundances in the measured air mixtures occur. Likewise, an existence of isotopologues has to be considered as well, like e.g. H_3O^+ (m/z 19) and $H_3O^+(H_2O)$ (m/z 37), whose natural isotopologues can also be detected on m/z 21 (ratio: $\frac{m/z19}{m/z21} \approx 500$) and m/z 39 (ratio: $\frac{m/z37}{m/z39} \approx 250$), respectively.

Quadrupole systems have a coarse mass resolution ($\frac{m}{\Delta m} \approx 200$) not allowing to resolve isobars (see also section 3.1.5, p. 50), which differ in mass by typically $0 - 0.2\,amu$. The analysis of superimposed signals on identical m/z therefore requires additional input in form of other chemical indicators or crosschecks using other instruments for classification. Nonetheless, a broad variety of organic compounds has been clearly assigned to corresponding masses for measurements using quadrupole systems. Tables are given e.g. in Blake *et al.* [2009]; de Gouw & Warneke [2007]; de Gouw [2003]; Karl [2003]; Warneke *et al.* [2003]; Yuan *et al.* [2014], and a selection is shown in table T 2.2 (p. 18).

4.1.1 Mission calibrations using pressurized test gas

During UBWOS 2012 bottle calibrations were performed every other day at the field site using the setup presented in figure F 4.1 (p. 59). One of the series, which was taken on 13 February 2012, is shown in figures F 4.2 (p. 60) and F 4.3 (p. 61). By evaluating the complete series of all performed calibration measurements within the mission period, universal mission calibration factors with appropriate error estimations have been derived. Table T 4.1 (p. 62) presents these factors.

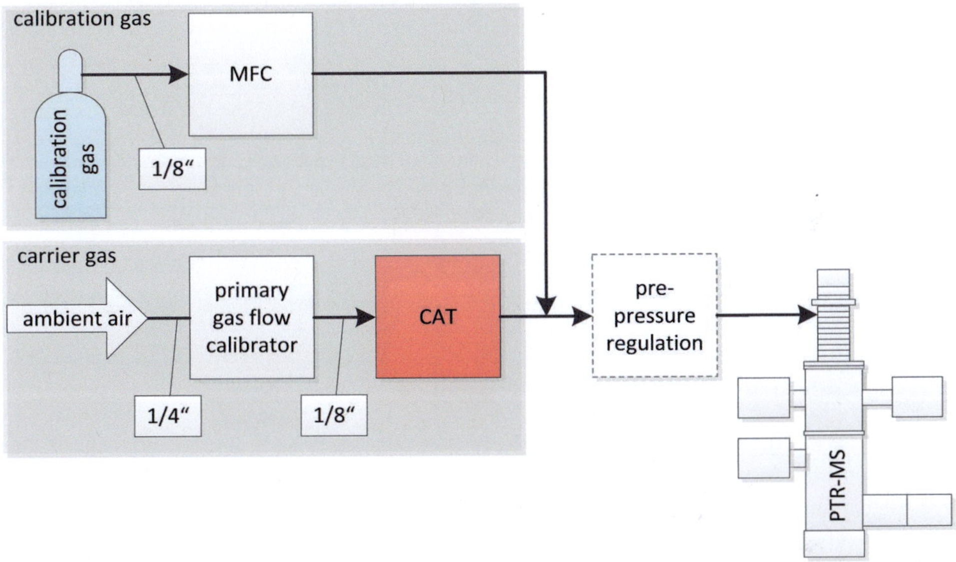

Figure F 4.1: *Setup for bottle calibration: The calibration gas originating from a gauged gas standard is controlled by a mass flow controller (MFC). It is intermixed to the carrier gas flow (ambient or zero air) before being injected to the pre-pressure regulation of the instrument. The carrier gas is controlled by a primary gas flow calibrator and is passed through the platinum catalyst (CAT) for cleansing.*

System accuracy The standard deviations (σ) of the sensitivities yielded on average $4.91\,\%$, which is within the expected fluctuation margin of calibration errors estimated with $\lesssim 5\,\%$. Nevertheless, it has to be emphasized that the actually observed deviations for the most compounds were clearly smaller, but the polar oxygenate methanol (m/z 33) with a σ of $16.97\,\%$ strongly contributed to this result.

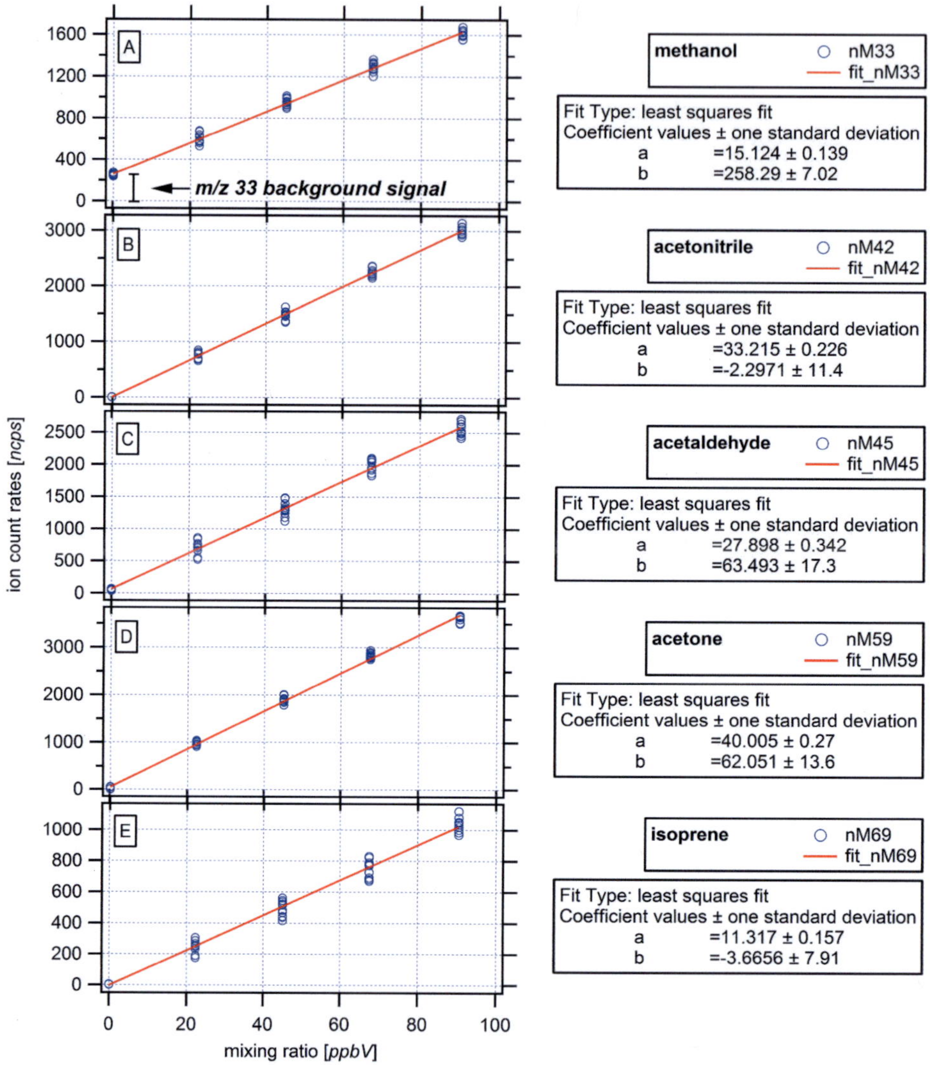

Figure F 4.2: *Bottle calibration performed on 13 February 2012: Normalized ion count rates (blue) and the least square fits (red) are plotted as mixing ratios in [ppbV] versus count rates in [ncps]. The fits ($y = a \cdot x + b$) are characterized by the slope a (→ calibration factor S) and the offset b (→ background signal). Presented compounds are* A *M33* ***(methanol)****,* B *M42* ***(acetonitrile)****,* C *M45* ***(acetaldehyde)****,* D *M59* ***(acetone)****,* E *M69* ***(isoprene)****.*

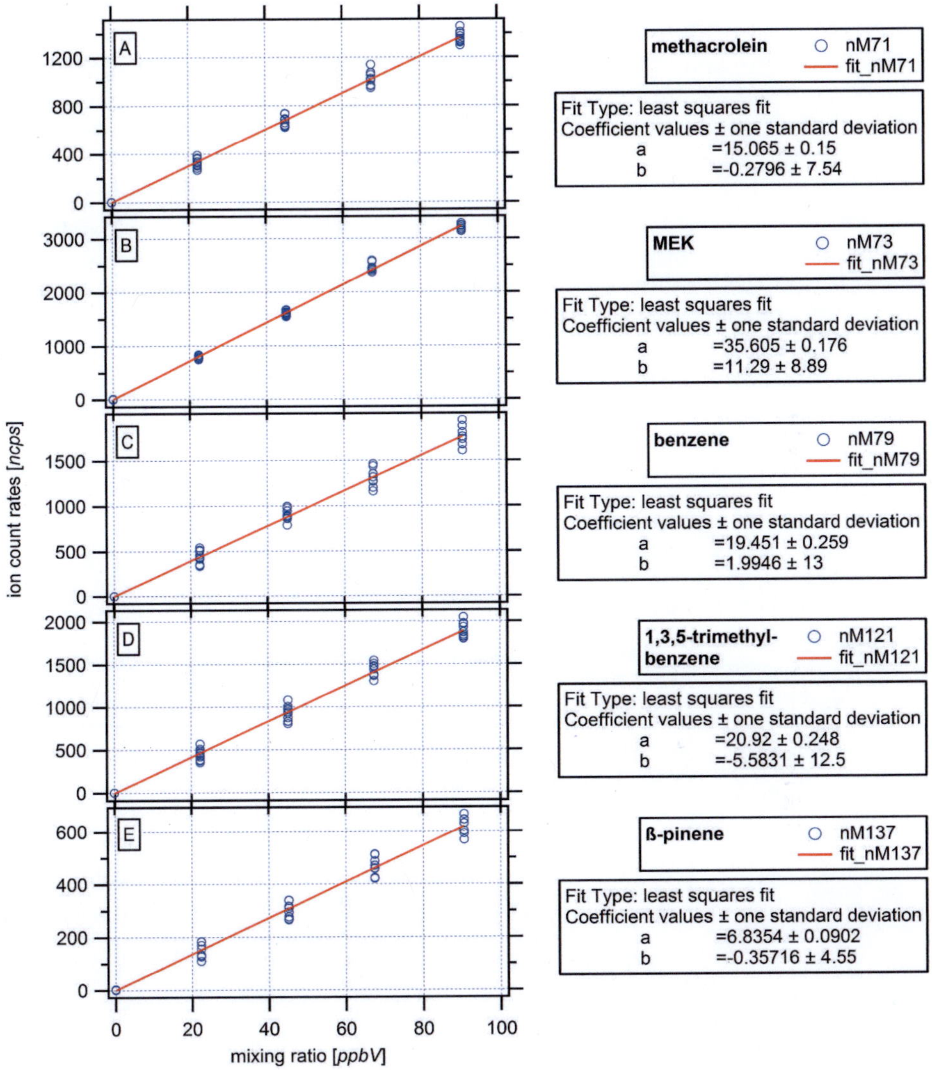

Figure F 4.3: *Bottle calibration performed on 13 February 2012: Normalized ion count rates (blue) and the least square fits (red) are plotted as mixing ratios in [ppbV] versus count rates in [ncps]. The fits ($y = a \cdot x + b$) are characterized by the slope a ($\rightarrow$ calibration factor S) and the offset b ($\rightarrow$ background signal). Presented compounds are* [A] *M71 (**methacrolein**),* [B] *M73 (**MEK**),* [C] *M79 (**benzene**),* [D] *M121 (**1,3,5-trimethylbenzene**),* [E] *M137 (β-**pinene**).*

These effects are likely to result from: ① the little mass leading to different mixing efficiencies during bottling compared to heavier compounds, ② effects due to an elevated water-sensitivity (see section 4.2, p. 72), ③ a strong interaction with metal surfaces in the applied MFC requiring a sufficient lead time (typically over night) in order to allow the amount of methanol in the diluted calibration standard to become stable [de Gouw & Warneke, 2007], and ④ a decrease of the methanol transferring efficiencies through tubing under very dry conditions by enhanced sticking to surfaces [Apel *et al.*, 2008]. Together with the normally stated standard uncertainty of $\sim 5\,\%$ for calibration gases and the estimated sum of $\sim 5\,\%$ for all additional uncertainties originating from the used instrumentation (e.g. MFCs), the total measurement accuracy for these measurements using a quadratic propagation of uncertainties (uncorrelated variables) can therefore be assumed to be:

$$accuracy[\%] \cong \sqrt{(error_{calibration})^2 + (error_{gasbottle})^2 + (error_{instrument})^2} \quad \text{(E 4.1)}$$

$$\cong \sqrt{(4.91\,\%)^2 + (5\,\%)^2 + (5\,\%)^2} \quad \text{(E 4.2)}$$

$$\cong 9\,\% \quad \text{(E 4.3)}$$

Compound	PTR-MS mass [amu]	Used X_R-factor	Calibration Factor S [$\frac{ncps}{ppbV}$]	Standard Deviation [$\frac{ncps}{ppbV}$]	Standard Deviation [%]
Methanol	33	0.5	15.14	2.57	16.97
Acetonitrile	42	0.65	32.16	1.63	5.07
Acetaldehyde	45	0.5	27.17	0.66	2.43
Acetone	59	0	40.25	1.57	3.90
Isoprene	69	0.5	11.36	0.21	1.85
Methacrolein	71	0.25	15.12	0.62	4.10
MEK	73	0.25	36.91	1.6	4.33
Benzene	79	0	19.14	0.65	3.40
1,3,5-Trimethyl-benzene	121	0.25	21.55	1.02	4.73
β-Pinene	137	0.5	6.94	0.16	2.31

Table T 4.1: *Calibration factors with error estimations derived for ULW-PTR-MS during UBWOS 2012 (averaged over the entire mission period Jan - Mar 2012) using the measured (by GC-MS) NOAA gas standard AAL071532. (for X_R-factor definition see section 2.2.3, p. 20)*

The derived accuracy of $\sim 9\,\%$ for the reduced selection of compounds reported here was well within the generally expected measurement accuracy of $\leq 20\,\%$ for the broad range of VOCs measured by PTR-MS [de Gouw & Warneke, 2007].

Adjustment of calibration factors S under mission-specific considerations using inter-system comparisons The adjustment of calibration factors for compounds not included in the used standard gas mixture was conducted by direct comparisons to the NOAA instruments. Their data uncertainties were stated with $15-20\,\%$ for hydrocarbons and $20-35\,\%$ for oxygenates measured by GC-MS and $20\,\%$ for reported compounds and $50\,\%$ for other masses measured by PTR-MS [de Gouw & Warneke, 2007; Gilman *et al.*, 2013]/(data specifications for UBWOS 2012[24]). Furthermore, despite the directly measured calibration factors S as listed in table T 4.1 (p. 62), the m/z assignment during UBWOS 2012 was subject to further mission-specific considerations. Crosschecks using NOAA GC-MS data and the exclusion of compounds, that did not exist in the unique chemical environment during the measurements indicated, that several observed compounds were not the compounds typically assigned with the respective m/z. Significant contribution to these signals therefore originated form other compounds, that are normally suppressed by the principal contributors. This particularly applied to m/z 69, which is normally associated with isoprene. Since the wide absence of isoprene emitting plants in the barren and predominantly desertic environment of the mission area, normally superposed small cycloalkanes were exclusively observed instead and the corresponding calibration factor was adapted. Likewise, the calibration factor for m/z 71, which is normally associated with MVK+MACR, was changed to a clearly smaller value in consequence of an observed interference with alkanes. The full list of measured m/z and derived calibration factors S for the measurements of ULW-PTR-MS during UBWOS 2012 is shown in table T 4.2 (p. 64). The corresponding system comparisons, that facilitated these adaptations, are discussed in detail in section 4.4 (p. 75).

4.1.2 Laboratory calibrations using a permeation tube and additional humidification - exemplified by formaldehyde (HCHO)

Humidity-sensitive calibrations The calibration of compounds with moisture sensitive calibration factors, which is due to their low ΔPA (see equation E 2.1, p. 16), is more complex, as appropriate humidification of the gas sample has to be considered in addition. This leads to non-linear correlations or calibration curves, respectively, between the detected signals in $[(n)cps]$ and the compounds' mixing ratios in $[ppbV]$ in dependence of the water contents of the sample air.

24 http://esrl.noaa.gov/csd/groups/csd7/measurements/2012ubwos/

	PTR-MS mass $[amu]$	Calibration factor S $[\frac{ncps}{ppbV}]$	Used X_R-factor	Compound	Comment
reported compounds	31	9	0	**Formaldehyde**	(ambient humidity)
	33	11.7	0.5	**Methanol**	
	42	31.81	0.65	**Acetonitrile**	
	45	34	0.5	**Acetaldehyde**	
	59	40.25	0	**Acetone**	
	73	35	0.25	**MEK**	
	79	17	0	**Benzene**	
	93	17	0.25	**Toluene**	
	107	17	0.25	sum of **C8-aromatics**	
	121	14	0.25	sum of **C_9-aromatics**	
	129	20	0.5	**Naphthalene**	
	135	20	0.5	sum of **C_{10}-aromatics**	
	137	6.94	0.5	**Monoterpenes**	
	149	20	0.5	sum of **C_{11}-aromatics**	
	163	20	0.5	sum of **C_{12}-aromatics**	
	177	20	0.5	sum of **C_{13}-aromatics**	
reported masses (not assigned to compounds)	28	20	0.5		
	30	20	0.5		
	41	20	0.5		
	43	1	0	alkanes‡	
	47	20	0		
	57	16	0.5	alkanes‡	
	61	20	0.5		
	69	0.4	0.5	cycloalkanes (alkenes$^{\top}$)	C_2-cyclohex.† (isoprene$^{\top}$)
	71	1.2	0.25	alkanes‡	cross corr. to alkanes
	81	40	0.5		
	83	0.8	0.5	cycloalkanes (alkenes$^{\top}$)	methylcyclopentane†
	85	30	0.5	alkanes‡	
	91	20	0.5		
	97	1.1	0	cycloalkanes (alkenes$^{\top}$)	methylcyclohexane†
	99	25	0.5		
	111	1.2	0.5	cycloalkanes (alkenes$^{\top}$)	C_2-cyclohexanes†
	125	20	0.5	cycloalkanes (alkenes$^{\top}$)	C_3-cyclohexanes†
	139	20	0.5		

Table T 4.2: *ULW-PTR-MS calibration factors S and used X_R-factors for reported compounds (first section; clear m/z assignment to compounds) and other masses (second section; masses with possible contributions from several compounds) as used during UBWOS 2012. († main cycloalkane; ‡ contributions from several alkanes (straight-chain and branched); $^{\top}$ at Horse Pool only small amount of alkenes detected (GC-MC); for X_R-factor definition see section 2.2.3, p. 20). The classification is in accordance with de Gouw & Warneke [2007]; Warneke* et al. *[2014]; Yuan* et al. *[2014].*

HCHO kinetics in the DT The formaldehyde (m/z 31) chemistry in the DT has been independently specified by Inomata *et al.* [2008]; Vlasenko *et al.* [2010]; Warneke *et al.* [2011b]. These studies served as basis for the presented calibration procedures.

The HCHO chemistry in the DT is particularly affected by the low proton affinity (PA(HCHO) = $712.9\,kJ\cdot mol^{-1}$), which necessitates paying attention to backward PTRs (with rate coefficient k_2) in addition to the normal forward reactions (with rate coefficient k_1):

$$H_3O^+ + HCHO \underset{k_2}{\overset{k_1}{\rightleftharpoons}} H{\cdot}HCHO^+ + H_2O \tag{R 4.20}$$

with:

k_1:	rate coefficient at 300 K	$1.4 \times 10^{-9}\,cm^3\cdot molecule^{-1}\cdot s^{-1}$
k_2:	rate coefficient at 300 K	$3 - 5 \times 10^{-11}\,cm^3\cdot molecule^{-1}\cdot s^{-1}$

[Hansel *et al.*, 1997; Vlasenko *et al.*, 2010]

Following Warneke *et al.* [2011b], the humidity-dependent H·HCHO$^+$ concentration in the DT is given by:

$$[H \cdot HCHO^+] = [H_3O^+] \cdot \frac{k_1 \cdot [HCHO] \cdot (1 - e^{-k_2\cdot[H_2O]\cdot\Delta t})}{k_2 \cdot [H_2O]} \tag{E 4.4}$$

where [HCHO], [H_3O^+], [H_2O] are the DT mixing ratios and Δt the reaction time. As [H_3O^+] is usually constant, primarily the humidity in the sample air, i.e. [H_2O], governs equation E 4.4 (p. 65) and thus the sensitivity for HCHO. Besides the thereof resulting non-linear calibration factorization, the non-negligible existence of back-reactions explains the lower efficiency of PTR compared to VOCs with a larger ΔPA.

Calibration setup with permeation tube The setup containing a heated water reservoir as humidification unit ($T_{humidification} = 25\,°C$) for the moisturization of the carrier gas is presented in figure F 4.4 (p. 66). In order to account for the effect of H_2O in the DT, the contributing water concentrations have to be known, as they are the sum of water vapor that is entering the DT with the calibration/sample gas flow and also that emanating from the IS [Vlasenko *et al.*, 2010]:

$$[H_2O]_{DT} = [H_2O]_{calibration/sample} + [H_2O]_{IS} \tag{E 4.5}$$

The H_2O mixing ratio in the DT is attuned using reaction R 4.20 (p. 65), which determines the measured distribution of the primary signals at m/z 19 (H_3O^+) and the water dimer at m/z 37 ($H_3O^+(H_2O)$) [Warneke *et al.*, 2001]. Thus, instead of the water mixing ratio the fraction $^{i(M37)}/_{i(M19)}$ is used to express [H_2O]$_{DT}$. The H_2O content in the sample or calibration/carrier gas flow can be estimated by measuring the absolute humidities (in [$\%RH$]) and adjusting for the pressure drop from ambient or calibration/carrier gas to DT conditions. During the laboratory measurement $^{i(M37)}/_{i(M19)}$ was varied between $1 - 40\,\%$ by means of changing carrier gas flows through the water reservoir (three-valve system). In the calibration gas flow, synthetic air passed over the permeation device containing paraformaldehyde housed in a Teflon cylinder (tube temperature $T_{permtube} = 40\,°C$, flow $F_{permtube} = 10.873\,sccm$). This produced a HCHO signal of $\sim 1.4\,ppm$ (determined by MOCCS), which has then been admixed to the carrier gas flow $F_{carriergas} = 89.905\,sccm$. Flows were controlled using MFCs and have been measured volumetrically (using a Bios DryCal Definer 220) for each humidity setting.

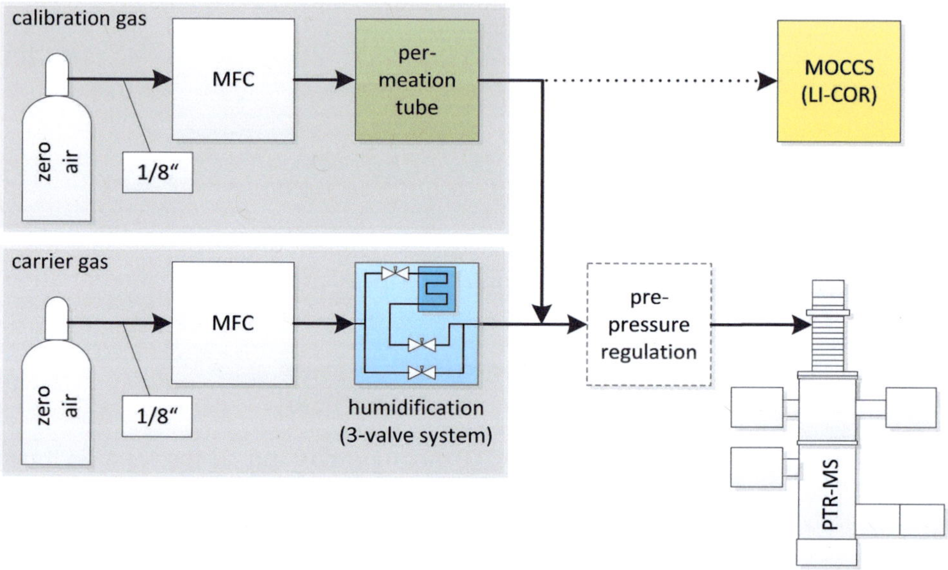

Figure F 4.4: *Calibration setup for compounds with humidity-dependent calibration factors using a permeation tube as calibration gas source: The calibration gas consisting of a zero air flow, that is controlled by a mass flow controller (MFC), is passed over the permeation tube (filled with solid/liquid calibration gas sources). It is then intermixed to the carrier gas flow or precisely gauged in the MOCCS system. The carrier gas, which is zero air controlled by a MFC, is humidified according to demand before being injected to the pre-pressure regulation of the instrument.*

Calibration curves Figure F 4.5 (p. 67) presents the derived calibration curves. The used DT parameters during this laboratory measurement were pressure $P_{drift} = 2.2\,hPa$ and potential difference $\Delta\Phi = 708\,V$. The deduced curves clearly illustrate the strong humidity dependence especially between a ${}^{i(M37)}/_{i(M19)}$ ratio of $2 - 10\,\%$.

During other PTR measurements in the context of an aircraft study, typically observed atmospheric ranges of the ratio ${}^{i(M37)}/_{i(M19)}$ have been reported to be between $\sim\ 2\,\%$ in the free troposphere and $\sim\ 14\,\%$ in the marine boundary layer

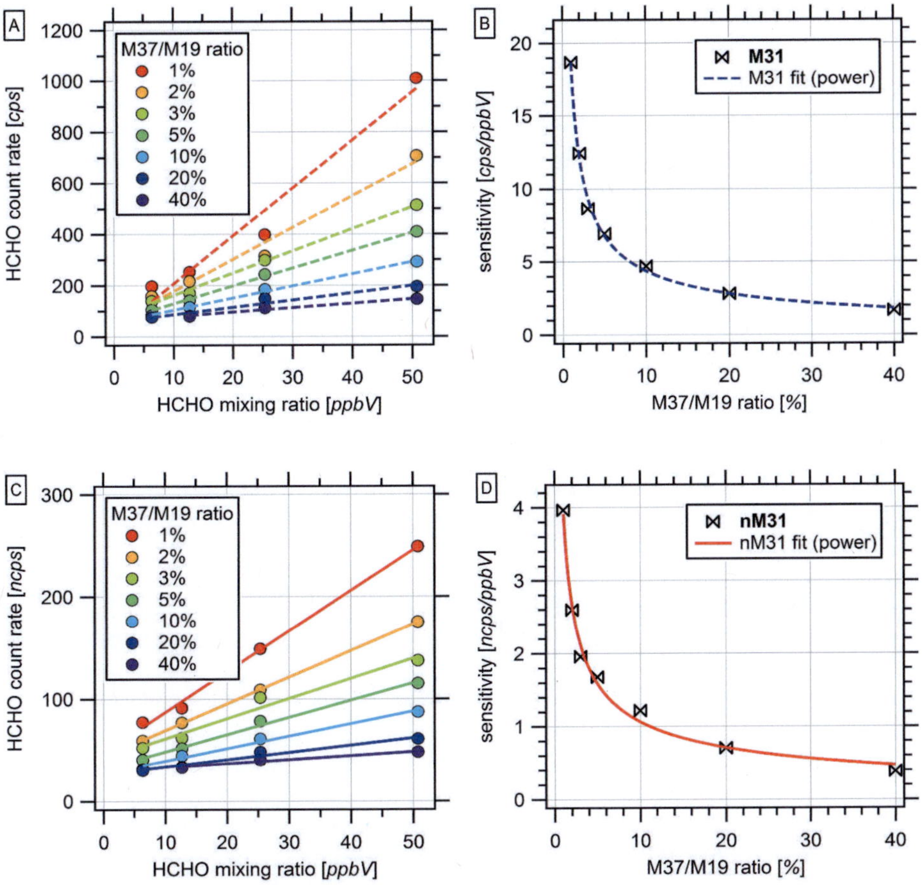

Figure F 4.5: *The derived calibration curves for formaldehyde (HCHO) using laboratory operating settings are shown for different humidities (represented by the ratio* $^{M37}/_{M19}$ *ranging between* $1 - 40\,\%$*). Panel* A *shows the HCHO mixing ratio in [ppbV] plotted versus m/z 31 ion count rates in [cps] and color-coded for the different humidities with each line representing a separate calibration. Panel* B *shows the humidity ratio plotted versus the calculated sensitivities (slopes from panel A) in* $[^{cps}/_{ppbV}]$*. Panels* C *and* D *represent corresponding normalized values.*

[Warneke *et al.*, 2011b]. The corresponding H_2O mixing ratios during these measurements were between $\sim 1 - 10\,g\cdot kg^{-1}$ for a ${}^{i(M37)}/_{i(M19)}$ ratio of $1 - 10\,\%$. However, it has to be emphasized that the differences in PTR-MS operating conditions, namely P_{DT}, voltage and temperature, only allow qualitative comparisons of the amount of water originating from the IS when comparing different instruments.

Furthermore, it is necessary to mention that these calibration curves base on laboratory measurements with differing instrumental settings compared to UBWOS 2012 and therefore cannot be compared with the corresponding HCHO sensitivity (at averaged humidity for ambient conditions) presented in table T 4.2 (p. 64).

Performance of HCHO measurements using PTR-MS Wisthaler *et al.* [2008] demonstrated in a comparative study (comprising different sensing principles like Hantzsch[25], DOAS[26], DNPH/HPLC[27]) at the atmosphere simulation chamber SAPHIR, that PTR-MS is well suited for reasonable on-line measurements of atmospheric formaldehyde concentrations in the sub-*ppb* range (at low absolute humidities $< 1\%$). Moreover, Jobson & McCoskey [2010] reported that additional drying of the air sample to a dew point of $-30\,°C$ (using a cryogenic trap to condensate and freeze H_2O vapor) is an option to significantly increase the low formaldehyde sensitivities by a factor of up to 7.

Nevertheless, as PTR-MS detection of formaldehyde at m/z 31 is strongly dependent on ambient H_2O concentrations and instrument settings, regular calibration is recommended. For future atmospheric HCHO measurements with ULW-PTR-MS tools need to be set up for stationary (/mobile) pre- and post-measurement (/online) calibrations using permeation devices with integrated dilution and humidification units. Likewise, simultaneous ambient H_2O data are necessary for deriving the correct sensitivity. Moreover, as the HCHO concentrations in the upper troposphere and lower stratosphere are typically small (around $100\,pptV$ in clean remote regions of the mid-troposphere; $20 - 30\,pptV$ in the upper troposphere [Fried *et al.*, 1999]), a further enhancement of the sensitivity by means of higher primary ion count rates is required to allow measurements with reasonable confidence levels of $\leq 20\,\%$.

4.1.3 Retroactive hydrogen sulfide (H_2S) calibration

PTR measurements of the non-VOC hydrogen sulfide The measurement of hydrogen sulfide (m/z 35) is particularly humidity-sensitive due to the low proton affinity of $PA(H_2S) = 705\,kJ\cdot mol^{-1}$. During UBWOS 2012 the ULW-PTR-MS measured H_2S experimentally, which was one of the first applications using PTR instruments in that context. Thus, the measurements initially started without calibration experiences. In order to though use the gained data set (further discussed in section

[25] Fluorometry based on the Hantzsch reaction
[26] Differential Optical Absorption Spectroscopy
[27] 2,4-dinitrophenylhydrazine derivatization followed by High Pressure Liquid Chromatography

7.6, p. 149), a retroactive calibration together with the deduction of the humidity-correlation for H_2S in the DT was necessary. Similar methods as shown for HCHO were applied (see section 4.1.2, p. 63). The calibration work was done and recently reported by Li *et al.* [2014].

H_3S^+ chemistry in the DT Both, the slightly exothermic PTR of H_3O^+ with H_2S (rate coefficient k_1) and the corresponding back reaction (rate coefficient k_2) with H_2O ($k_1 \gg k_2$) have to be considered as they apply to ambient humidity conditions [Li *et al.*, 2014]:

$$H_3O^+ + H_2S \underset{k_2}{\overset{k_1}{\rightleftharpoons}} H_3S^+ + H_2O \qquad \text{(R 4.21)}$$

with:

k_1:	rate coefficient at 298 K	$1.9 \times 10^{-9}\ cm^3 \cdot molecule^{-1} \cdot s^{-1}$ [Tanaka *et al.*, 1978]
k_2:	rate coefficient at 298 K	$4.4 \times 10^{-12}\ cm^3 \cdot molecule^{-1} \cdot s^{-1}$ [Tanaka *et al.*, 1978]

According to analogous approaches for hydrogen cyanide (HCN) (PA(HCN) = $712.9\ kJ \cdot mol^{-1}$) presented by Knighton *et al.* [2009] and with the help of equation E 2.11 discussed in section 2.2.3 (p. 20), the DT concentration of H_3S^+ ions is given by:

$$[H_3S^+] = [H_3O^+] \cdot \frac{k_1 \cdot [H_2S] \cdot (1 - e^{-k_2 \cdot [H_2O] \cdot \Delta t})}{k_2 \cdot [H_2O]} \qquad \text{(E 4.6)}$$

where $[H_2S]$, $[H_2O]$, $[H_3O^+]$ are the corresponding DT mixing ratios and Δt the reaction time. With the normally higher DT mixing ratio of H_2O (typically 1%) than H_2S ($< 10ppbV$) ($[H_2O] \gg [H_2S]$), the exponential term can be neglected and only the ratio ${}^{(k_1 \cdot [H_2S])}/_{(k_2 \cdot [H_2O])}$ determines the final $[H_3S^+]$ in the DT. As the ion kinetic energy is elevated in the DT, the endothermic reactions (i.e. k_2) may be more important than simply due to their corresponding reaction enthalpies. Similar to HCHO, the back reactions decrease the efficiency of the production of protonated H_2S, which is therefore much less efficient than the production of most protonated VOCs. According to equation E 4.6 (p. 69) $[H_3S^+]$ is determined by the humidity $[H_2O]$ and the reaction time Δt as shown in figure F 4.6 (p. 71).With typical operating settings (attached in table T G.2, p. 177) the resulting reaction time is $\sim 100\,\mu s$ (cp. with illustration). It is worth noting, that the forward and backward reactions are not necessarily in equilibrium. As shown in figure F 4.6 (p. 71), low H_2O concentration conditions lead to residence times of H_3S^+ ions in the DT that are insufficient for backward reactions to get into an equilibrium with forward reactions (at typical settings). The consequence then is an enhanced production of H_3S^+ ions at lower water concentrations.

Deduction of H_2S calibration parameters Instrumental field/laboratory comparisons between the NOAA PTR-MS and the PTR-ToF-MS from the University of

Wyoming in Laramie, Wyoming/USA (Department of Atmospheric Sciences[28]) were performed by Li *et al.* [2014]. The instrument settings of the involved systems were typical for PTR measurements and are attached in table T G.2 (p. 177). Due to their influences to the PTR processes, these parameters (DT pressures, electrical potentials, internal temperatures and IS water vapor flows) and the behaviors of other compounds used as proxies (aromatics, oxygenated VOCs) were compared for all performed studies in order to assess the general instrumental properties of the instruments that rate humidity dependent measurements.

- Data from the NOAA PTR-MS and the PTR-ToF-MS were used to explore the detection of H_2S with regard to the strong humidity dependence in direct comparison to each other. The derived H_2S sensitivity of the NOAA PTR-MS was determined to be $0.6 - 1.4\,{}^{ncps}/_{ppbV}$ ($20 - 34\,{}^{cps}/_{ppbV}$) ($H_2O$ mixing ratios between $1 - 6\,g{\cdot}\,kg^{-1}$) during UBWOS 2013, which is about $3 - 10\,\%$ of most other compounds detected by PTR-MS. Reliable hydrogen sulfide detections down to levels of $0.35\,ppbV$ (signal-to-noise = 3) have been reported for ambient conditions (H_2O mixing ratio: $2.8\,g{\cdot}\,kg^{-1}$ on average).
- Laboratory test series with the same instruments have been performed using a H_2S calibration cylinder as source (concentration levels between $41 - 100\,ppbV$ H_2S). The humidity was changed by dilution with differently saturated zero air mixtures (H_2O mixing ratios between $0 - 2.61\,\%$). Resulting sensitivities for the NOAA PTR-MS and PTR-ToF-MS are attached in figure F G.1 (p. 178). The marked field calibration of the NOAA PTR-MS at ambient UBWOS 2013 conditions showed the sensitivity of $1.04\,{}^{ncps}/_{ppbV}$, that agreed within $20\,\%$ with the laboratory calibration results.

Concurrent H_2S measurements using a Picarro cavity ring down spectroscopy (CRDS) instrument for H_2S/CH_4 measurements have been used to verify the measured mixing ratios. Further details of the calibration factor deduction are attached to appendix G (p. 177).

H_2S calibration of the ULW-PTR-MS Figure F 4.7 (p. 71) illustrates the comparisons of instrumental sensitivities between NOAA PTR-MS and ULW-PTR-MS using correlations of selected standard compounds. The applied linear fit shows a slope of 0.99 ± 0.05 ($R^2 = 0.9$). Despite different instrument settings, the sensitivities agreed well for a wide range of compounds within their uncertainties (accuracy $30\,\%$) and thus enable a transfer of the H_2S calibration factor to ULW-PTR-MS using that of the NOAA PTR-MS. The estimated value from the fit is determined with $1.03 \pm 0.05\,{}^{ncps}/_{ppbV}$. The humidity conditions during UBWOS 2012 were similar to those for the laboratory calibrations, thus indicating that the estimated factor is a reasonable approximation. The little deviations from the 1:1 line were presumably due to different instrument settings, that affected the (mass-dependent) PTR efficiencies and the instrumental transmission factors $T_{H_2S/VOCs}$.

[28] http://www.uwyo.edu/atsc/

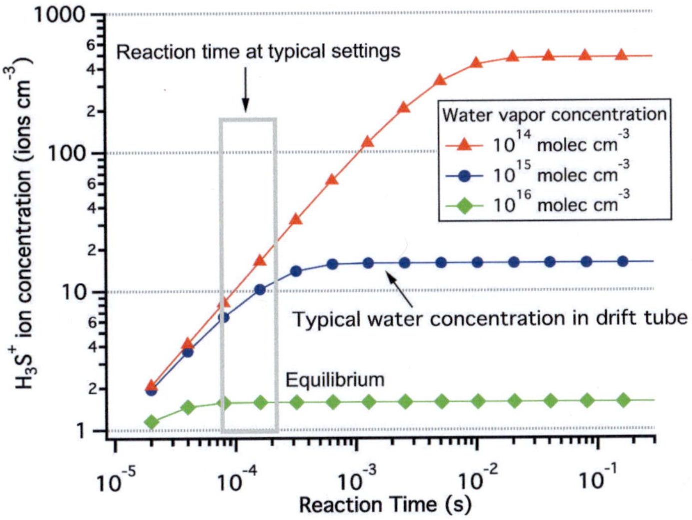

Figure F 4.6: *Concentration of formed* H_3S^+ *ions calculated for different humidities* (0.5×10^{15} *to* 2×10^{15} *molecules per* cm^3 */ equivalent to* H_2O *mixing ratios of* $0.9\,\%$ *to* $4.7\,\%$*) and plotted as function of the reaction time. (adopted from Li* et al. *[2014])*

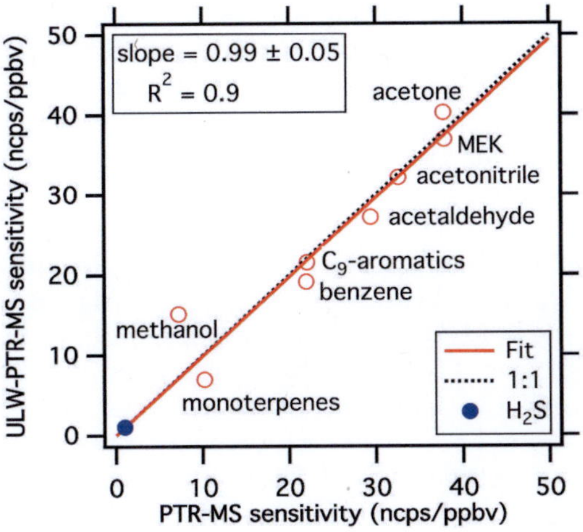

Figure F 4.7: *Calibration factors in* $[^{ncps}/_{ppbV}]$ *of selected compounds (red circles) from the NOAA PTR-MS calibration (x-axis) and the ULW-PTR-MS calibration (y-axis) with linear fit line (red). The estimate of ULW-PTR-MS calibration factor for* H_2S *(blue dot) from the NOAA PTR-MS calibration at the field site during UB-WOS 2013, based on the comparison of calibration results for the other compounds between the two PTR-MS instruments. (adapted from Li* et al. *[2014])*

4.2 Interpretation of ion processes - m/z 33 ambiguity

Ion formation or fragmentation processes below the DT do bias the initially ionized trace gas compositions. A prime example for that effect is at m/z 33, where isobaric species are detected. The result of that is visible in form of a disproportionately high background signal (chemical noise) when calibrating for methanol (CH_4O; protonated: CH_5O^+; m/z 33) using standard instrumental settings (see panel A in figure F 4.2, p. 60). Methanol is a comparatively light compound with slight humidity-dependency due to its small proton affinity $PA_{methanol} = 754.3\,kJ \cdot mol^{-1}$. Besides CH_5O^+ (measured at m/z 33.0335) two other isobaric compounds representing instrumental background peaks contribute to this m/z 33, which are m/z 32.9957 corresponding to ($O^{18}O^+$ + O_2H^+) and m/z 33.0204 corresponding to $DNOH^+$ [Müller *et al.*, 2010]. Due to restrictions in the mass resolution of the quadrupole system (unit mass resolution) not allowing an effective m/z separation of the contributing peaks, the quoted investigations have been performed using a PTR-ToF-MS instrument (mass resolving power of about $\frac{m}{\Delta m} = 5'000$) [Graus *et al.*, 2010]. The two contributors to the m/z 32.9957 signal as significant peak have been found to be formed during different processes: ① $^{16}O^{17}O^+$ (associated with processes in the IS) and ② O_2H^+ (associated with processes in the extraction region). An endothermic PTR from hydronium ions to O_2 according:

$$H_3O^+ + O_2 \longrightarrow H_2O + O_2H^+ \qquad \text{(R 4.22)}$$

is therefore indicated, as the collision energies behind the NC plate are high, whereas P and with that the collision frequencies are reduced. The fraction of formed O_2H^+ ions is strongly dependent on the physical conditions like the pressures P_{IVC}, $P_{quadrupole}$ and ion velocities ν_{ion} around the extraction lenses (i.e. between DT and following high-vacuum chambers with NC plate and ion optics). With higher extraction voltages U_{bottom}/U_{NC} and $U_{NC}/U_{ion-optics}$ the general sensitivities together with the enhanced formation of m/z 33 background signals strongly increase. Lower extraction voltages reduce the efficiency of O_2H^+ formation and at the same time increase the possibility for further collisions by means of fragmentation/loss of H^+. In parallel to that, an increase of H_2O molecules in the sample air decreases the O_2^+ ions coming from the IS (O_2^+ and H_2O react together in a multi-step reaction [de Gouw & Warneke, 2007; Good, 1970]) and therewith enhances the possibility for further fragmentation processes of O_2H^+. The result is a superposition of both effects making the m/z 33 signal particularly sensitive to moisture (humidity-dependency of ① methanol and ② the underlying ion chemistry of the background signals) and susceptible to manipulation by slight changes of the electrical potentials at the extraction system. Having regard to the presented high background signals on m/z 33, a possible reduction of these effects by fine-tuning of the extraction voltages in the intermediate chambers is recommended for future measurements. Likewise, the diminution of the recorded mass range around m/z 33 is suitable in order to suppress the occurrence of disturbing $^{16}O_2^+$ ions within the methanol signals and should be considered as well.

4.3 Limit of Detection (LOD) - exemplified by acetone

The limit of detection of PTR systems (introduced in detail in paragraph 'Limit of Detection (LOD)' in section 2.2.3, p. 24) is the mixing ratio, where $\Delta VMR/VMR = 33\,\%$ (signal to noise ratio of 3). It is calculated for the time series presented in figure F 4.8 (p. 74). The data comprise continuous measurements in a two-week period during UBWOS 2012 (31 January - 14 February 2012) (with dwell times $\tau = 1\,s$). Shown acetone (m/z 59) data consist of ambient air signals with an averaged $I_{m/z59,amb} = 284\,cps$ as well as of scrubbed background signals with an averaged $I_{m/z59,bgd} = 91\,cps$ and are complemented by the primary H_3O^+ (m/z 19) signals with an averaged $I_{H_3O^+} = 4.27 \times 10^6\,cps$. The determined acetone calibration factor $S_{acetone} = 40\,^{ncps}/_{ppbV}$ (slope a in panel D of figure F 4.2, p. 60) is used for the calculation of the average $VMR_{acetone} = 1132\,pptV$ with equation E 2.28 (p. 25) and the average $\Delta VMR_{acetone} = 113\,pptV$ with equation E 2.29 (p. 25). The LOD is calculated using equation E 2.27 (p. 24) and results in $LOD_{acetone} \simeq 168\,pptV$. Comparable values reported in the literature are $84\,pptV$, but with dwell times of $\tau = 5\,s$ [de Gouw, 2003; Warneke *et al.*, 2005].

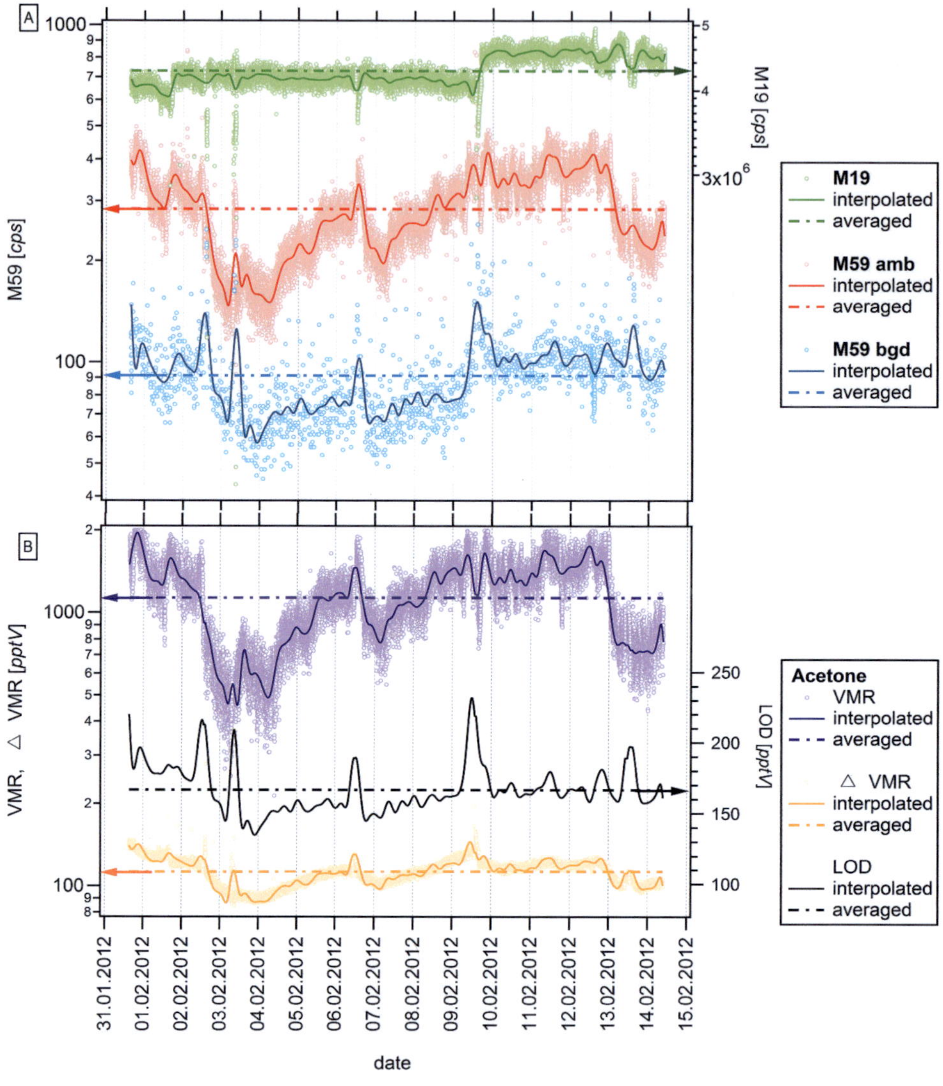

Figure F 4.8: *Acetone (m/z 59) time series measured during UBWOS 2012 for two weeks. Data are shown in raw counts (color-coded circles), as interpolated line using a pre-average of 100 nodes (solid lines) and in form of averaged values (dashed lines).* A *presents m/z 19 and m/z 59 ambient/background signals in [cps].* B *presents the post-processed VMR and ΔVMR (using* $S_{acetone} = 40\,^{ncps}/_{ppbV}$*) and the LOD (using the interpolated M21 and M59 background signals) in [pptV].*

4.4 Performance assessment by comparisons to other systems

The evaluation comprises a reference period of several weeks during UBWOS 2012 allowing the long-run analysis of the ULW-PTR-MS by comparing to the NOAA PTR-MS [de Gouw & Warneke, 2007] and the NOAA GC-MS [Gilman *et al.*, 2010]. The mean general sensitivity of the ULW-PTR-MS at that time was around $100 - 200\,{}^{cps}/_{ppbV}$ and that of the NOAA PTR-MS around $500 - 1000\,{}^{cps}/_{ppbV}$ [Warneke *et al.*, 2014]. Table T 4.3 (p. 75) summarizes the duty cycles and time bases of the used data sets. The contrasting of the two PTR systems was done using $1\,min$-averages. For all three instruments the contrasting using $5\,min$-averages was restrained to the corresponding GC integration periods every half hour. Calibration equipment was shared between both PTR instruments. The two separate inlet lines (① NOAA PTR-MS; ② NOAA GC-MS + ULW-PTR-MS) were heated.

Instrument	Measuring cycle	Averaged to	
•ULW-PTR-MS	· slow: $39\,sec \cdot cycle^{-1}$ · fast: $19\,sec \cdot cycle^{-1}$	$\cdot 1\,min/ \cdot 5\,min$	(37 masses)
•NOAA PTR-MS	$\cdot\, 27\,sec \cdot cycle^{-1}$	$\cdot 1\,min/ \cdot 5\,min$	(31 reported masses)
•NOAA GC-MS	· $5\,min$ integration (every $30\,min$)	$\cdot 5\,min$	(65 reported masses)

Table T 4.3: *Used data sets for the instrumental comparisons with duty cycles of the instruments and applied time bases.*

Three-instruments-comparisons In figure F 4.9 (p. 77) a representative 7-hour interval (recorded on 9 February 2012; 07:00 - 14:00 MST) of methanol (m/z 33) mixing ratios is presented. The first panel shows the time series for the diverse instruments and integration times, the second the relative deviations Δmethanol in [%] of the ULW-PTR-MS value and the third the prevailing wind conditions measured at the Horse Pool walk-up tower in $19\,m$ above ground.

In general, good reproductions of fine structures with both PTR instruments on the $1\,min$ base were ascertained, like e.g. 08:30 - 12:00 MST. Likewise, good agreement of ULW-PTR-MS and GC-MS integral data was observed. The reason for partially diverging NOAA PTR-MS data from that (e.g. 07:00 - 08:00; 12:00 MST) is due to the measuring setup with sampling probes that were located few meters apart from each other and without perfectly coordinated inlet line lengths. Thus, it is likely that different air parcels with individual compositions or with little time lags, respectively, have been analyzed during these periods, since the measurement region was characterized by random emissions from countless point sources and turbulent wind fields on site. No specific day-night differences or other interferences have been observed. The best agreements have been seen during periods with mostly stable winds in speed and direction, like e.g. 08:30 - 11:30 MST.

Notwithstanding that a good general consistency for all three VOC instruments has been determined (within their uncertainties), the described biasing effects are symptomatic for the kind of performed ambient air comparisons and have to be kept in mind for the further analysis.

Correlations of 1:1 reference data Scatter plots of selected data from one instrument vs. the corresponding data from another instrument are used in order to evaluate the performance of the ULW-PTR-MS. The correlations are specified using linear fits (least squares regression). Figure F 4.10 (p. 78) shows the correlation plots for the two PTR instruments. The observed enhanced scattering is possibly due to the setup with separated sampling probes. As particularly aromatic compounds (benzene, toluene, C_8-, C_9-aromatics) were affected by that, which were randomly injected to the sampled air from various point sources around (fossil fuel production area), the varying values can be explained by the individual compositions of different air parcels in periods with changing winds (see figure F 4.9, p. 77). The same time series of the PTR instruments are presented in figure F 4.11 (p. 79) in consideration of the GC-time pattern. Through the changing from $1\,min$-averages to $5\,min$-averages the agreements between both series clearly enhanced, as especially the impacts due to wind fields and other runaway values were compensated. The contrasting of ULW-PTR-MS vs. NOAA GC-MS presented in figure F 4.12 (p. 80) agreed best, which is most likely due to the setup consisting of the jointly used inlet line. An approximately perfect consistency with a coefficient of determination of $R^2 = 0.998$ has been achieved for methanol. The correlation results are summarized in table T 4.4 (p. 81).

Conclusions for the ULW-PTR-MS The performance assessment of ULW-PTR-MS, that was facilitated by the unique situation of three high-sensitivity VOC instruments directly measuring side by side, attested the system's high accuracy. The results of this characterization are as good as it can be expected for ambient air comparisons under field conditions. The instruments generally agreed within $\sim 10\,\%$ for all reported VOCs, which gives confidence for the use of data that were obtained in the context of the field mission UBWOS 2012.

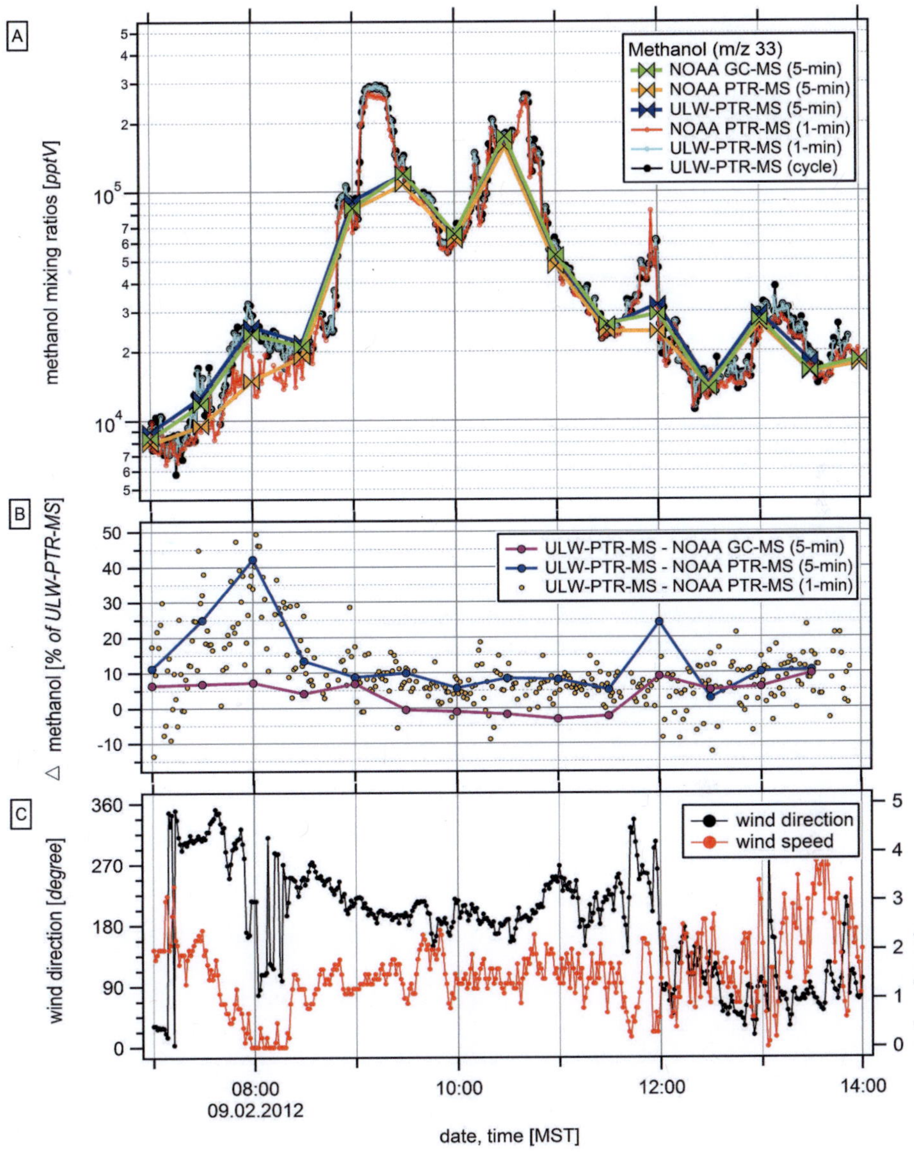

Figure F 4.9: *The methanol (m/z 33) time series measured on 9 February 2012 from 07:00 - 14:00 MST at the Horse Pool field site using NOAA GC-MS, NOAA PTR-MS and ULW-PTR-MS. Panel* A *shows the methanol mixing ratios in [pptV] and panel* B *the differences between the values in [%] of the ULW-PTR-MS value. Panel* C *shows the wind direction in [degree] and speed in [$m \cdot s^{-1}$] (1 min-averages).*

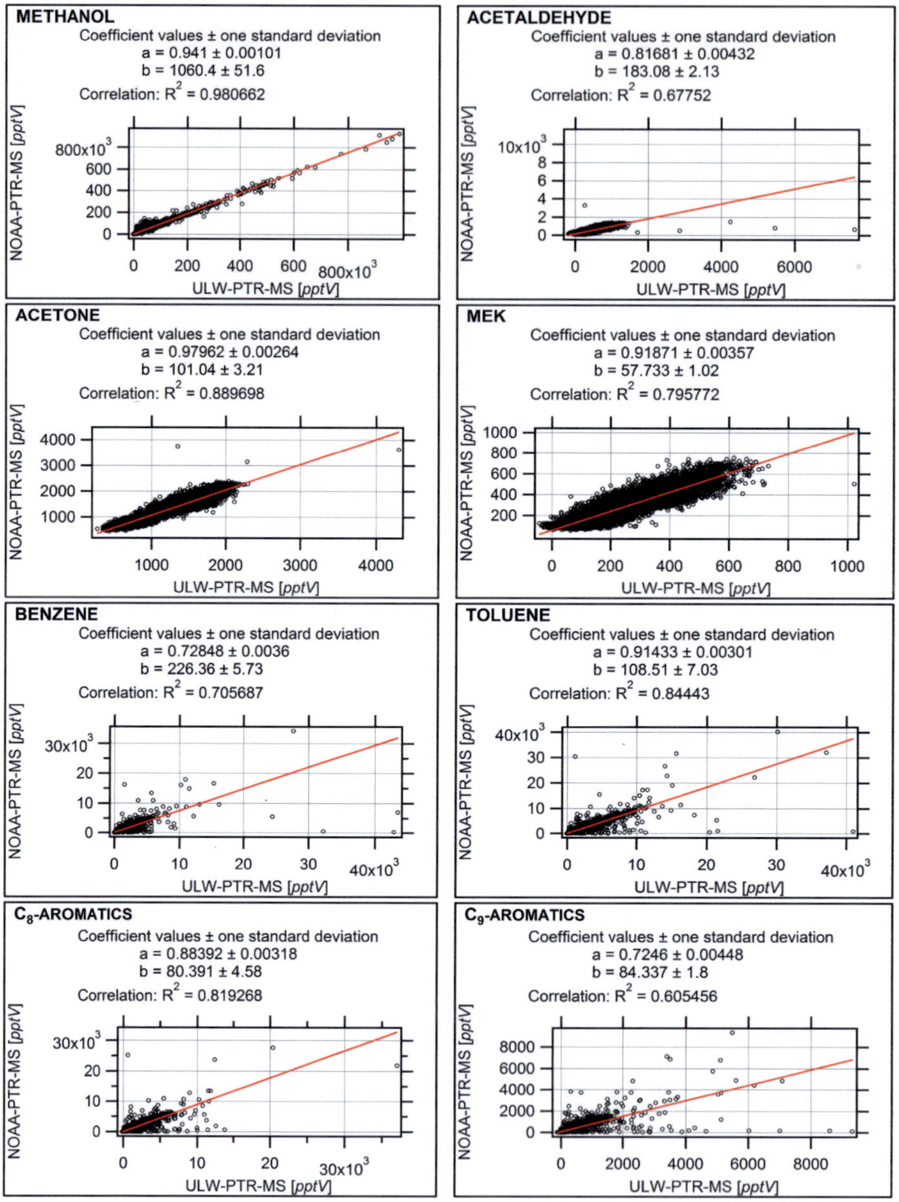

Figure F 4.10: *Correlation plots of* **ULW-PTR-MS vs. NOAA PTR-MS** *data on the* <u>*mission* 1 min *time frame*</u> *with slopes a, offsets b and coefficients of determination R^2 for ambient air compounds (methanol, acetaldehyde, acetone, MEK, benzene, toluene, C_8-aromatics, C_9-aromatics) measured for a 3 week period during the mission UBWOS 2012.*

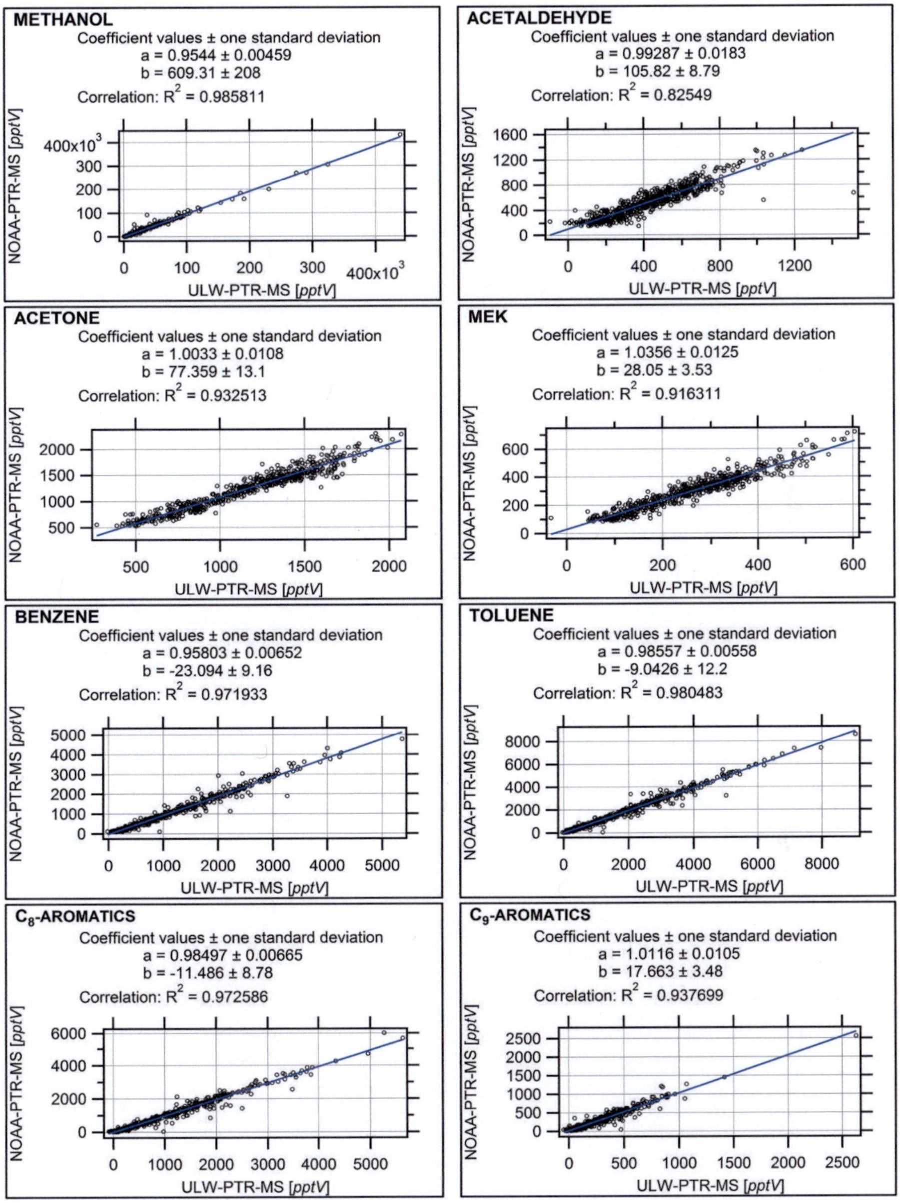

Figure F 4.11: *Correlation plots of* ***ULW-PTR-MS vs. NOAA PTR-MS*** *data on the* *GC-MS* $5\,min$ *time frame with slopes* a*, offsets* b *and coefficients of determination* R^2 *for ambient air compounds (methanol, acetaldehyde, acetone, MEK, benzene, toluene,* C_8*-aromatics,* C_9*-aromatics) measured for a 3 week period during the mission UBWOS 2012.*

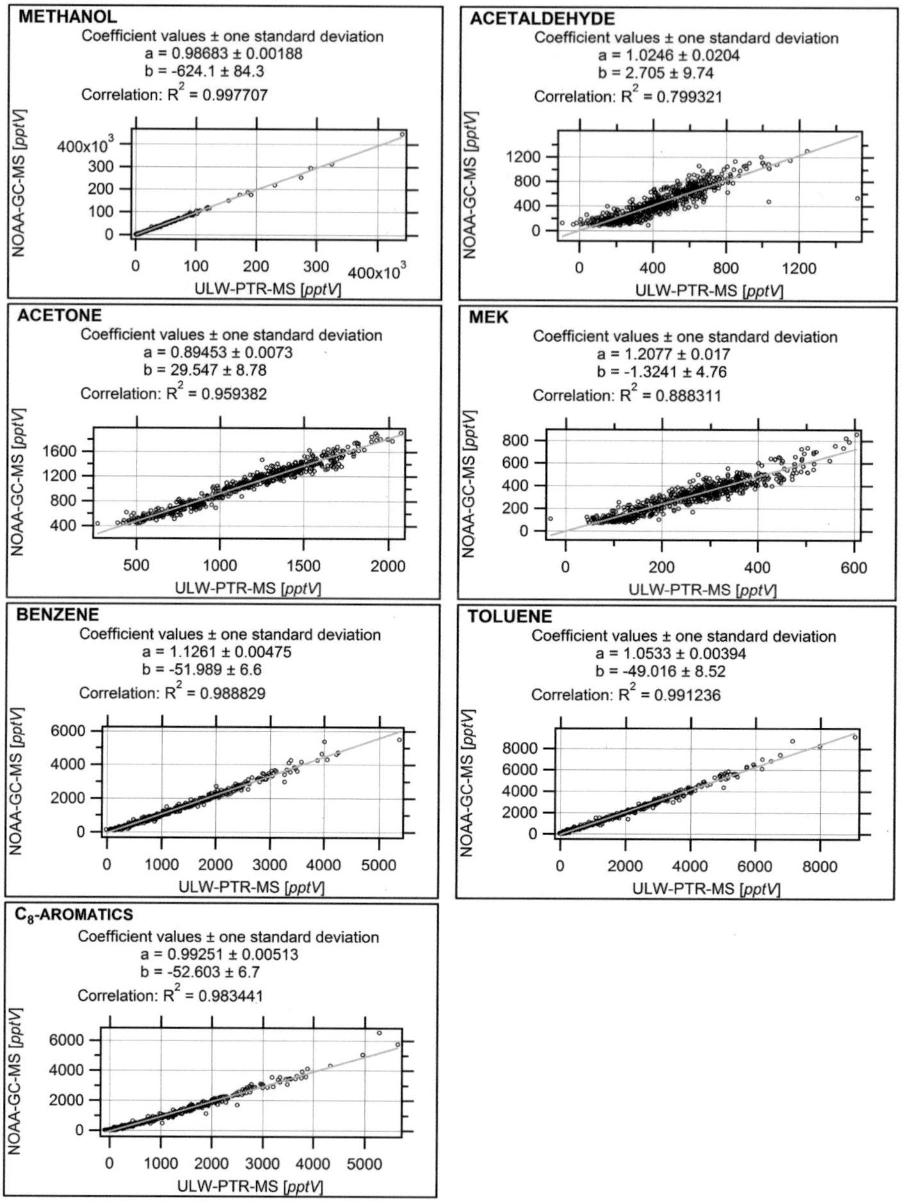

Figure F 4.12: *Correlation plots of* ***ULW-PTR-MS vs. NOAA GC-MS*** *data on the GC-MS 5 min time frame with slopes a, offsets b and coefficients of determination R^2 for ambient air compounds (methanol, acetaldehyde, acetone, MEK, benzene, toluene, C_8-aromatics) measured for a 3 week period during the mission UBWOS 2012.*

		NOAA PTR-MS		NOAA PTR-MS		NOAA GC-MS	
		mission time 1 min-averages		GC time 5 min-averages		GC time 5 min-averages	
		a	R^2	a	R^2	a	R^2
ULW-PTR-MS	**Methanol**	0.941 ±0.001	0.981	0.954 ±0.005	0.986	0.987 ±0.002	0.998
	Acetaldehyde	0.817 ±0.004	0.678	0.993 ±0.018	0.825	1.025 ±0.020	0.799
	Acetone	0.980 ±0.003	0.890	1.003 ±0.011	0.933	0.895 ±0.007	0.959
	MEK	0.919 ±0.004	0.796	1.036 ±0.013	0.916	1.210 ±0.017	0.888
	Benzene	0.728 ±0.004	0.706	0.958 ±0.007	0.972	1.126 ±0.005	0.989
	Toluene	0.914 ±0.003	0.844	0.986 ±0.006	0.980	1.053 ±0.004	0.991
	C_8-aromatics	0.884 ±0.003	0.819	0.985 ±0.007	0.973	0.993 ±0.005	0.983
	C_9-aromatics	0.725 ±0.004	0.605	1.012 ±0.011	0.938	n/a	

Table T 4.4: *The summary of derived correlation parameters originating from the contrastings according to correlation plots F 4.10 (p. 78), F 4.11 (p. 79), F 4.12 (p. 80), with a being the slopes and R^2 the coefficients of determination.*

Chapter 5

Future Applications

The system ULW-PTR-MS has been developed to be deployed on board of aircraft. The instrumental prototype ULW-PTR-MS, which is customized for the mission OMO aboard the HALO aircraft (see section 5.2, p. 86), is supplemented by a sister instrument that is planned to replace the present CARIBIC mass spectrometer (installed in 2005) in a major turnover in spring 2015 (see section 5.1, p. 83).

5.1 Regular applications aboard a Lufthansa passenger aircraft - CARIBIC

In CARIBIC[29] (Civil Aircraft for the Regular Investigation of the atmosphere Based on an Instrument Container) trace gases, water vapor, ice crystals and aerosols are globally monitored using a fully automatic instrument container, which is monthly deployed on board a Lufthansa Airbus A340-600 passenger aircraft (named: D-AIHE 'Leverkusen') for four to six consecutive long-duration flights between Germany (Frankfurt a. Main, Munich) and various destinations around the globe. The scientific instrumentation in the container consists of 15 systems for measurements of altogether ~ 100 trace gases and aerosol parameters. The project is briefly presented in figure F 5.1 (p. 85) in form of some portrayed key characteristics. Further details and system descriptions of the CARIBIC program can be found in Brenninkmeijer *et al.* [1999, 2007].

As worldwide monitoring project CARIBIC is operational since 1997 (initally using a Boeing 767 by LTU and since 2005 an Airbus A340-600 by Lufthansa) and provides an unique global data set covering on important species concerning climate change and air quality. Since the normal cruise altitudes of the used passenger aircraft mainly range between $9 - 12\,km$, especially the UT/LMS region (Upper Troposphere and LowerMost Stratosphere) is the main focus of attention. Thereby multiple crossings of the tropopause as intermediate boundary layer between UT and LMS (further defined in Zahn & Brenninkmeijer [2003]) allow conclusions concerning atmospheric chemistry and dynamics in the UT/LMS, which is a region

[29] `http://www.caribic-atmospheric.com/`

showing key influences onto weather and climate [Gettelman *et al.*, 2011; Randel & Jensen, 2013].

Since 2012 CARIBIC is part of the large-scale European Research Infrastructure IAGOS[30] (In-service Aircraft for a Global Observing System) and will be coordinated and operated by IMK[31] (at KIT in Karlsruhe/Germany) as of 2015.

CARIBIC PTR-MS - deployment/tasks The CARIBIC PTR-MS (version I) belongs to the base instrumentation since the relaunch of the project in 2005 on the Lufthansa carrier and has been used so far to determine acetone (m/z 59; C_3H_6O) (available data for > 180 flights), methanol (m/z 33; CH_4O) and acetonitrile (m/z 42; C_2H_3N) concentrations. These investigations are of interest as VOCs, like e.g. acetone, are directly involved in the production of HO_x radicals (see section 1.2.2, p. 8), which then again affect the degradation of greenhouse gases (e.g. methane) and air pollutants. The instrument and relevant implications of measured acetone concentrations are described in Neumaier *et al.* [2014]; Sprung & Zahn [2010]. Likewise, acetonitrile as clear tracer for air affected by biomass/biofuel burning events [Andreae *et al.*, 2001; Dunne *et al.*, 2012; Sprung *et al.*, 2001; Warneke *et al.*, 2011a] is used during CARIBIC to evaluate appropriate occurrences on global scales, like e.g. fire plumes over Asia in Baker *et al.* [2014]; Lai *et al.* [2010]. As example of the scientific coverage under direct consideration of modeled ECMWF[32] 5-day backward trajectories [Scheele *et al.*, 1996], the monitoring of pollution over Asia, as discussed by Baker *et al.* [2014], or the computation of vertical (across the tropopause) and horizontal climatologies in consideration of seasonal cycles, as discussed by Elias *et al.* [2011]; Neumaier *et al.* [2014]; Sprung & Zahn [2010], may be quoted.

As part of a scheduled revision of the instrument container in spring 2015, the present CARIBIC PTR-MS (version I) is planned to be replaced by the newer setup (version II) in order to upgrade its performance and reliability and to decrease the payload weight. This future instrument CARIBIC PTR-MS II will be similar to ULW-PTR-MS, which simplifies their maintenance, operations and quality management by means of direct transfer of experiences gained on board of different carriers. It uses the same instrumental body adapted to the platform-dependent CARIBIC aircraft rack and fitting to the missions' unique infrastructural conditions in the used instrument container. While being constructed this following instrument has already been engaged to perform instrumental testings during research flights in the context of the mission MUSICA 2013 (see appendix D, p. 169).

[30] `http://www.iagos.org/`

[31] Institute for Meteorology and Climate Research; `http://www.imk-asf.kit.edu/`

[32] European Centre for Medium-Range Weather Forecasts; `http://www.ecmwf.int/`

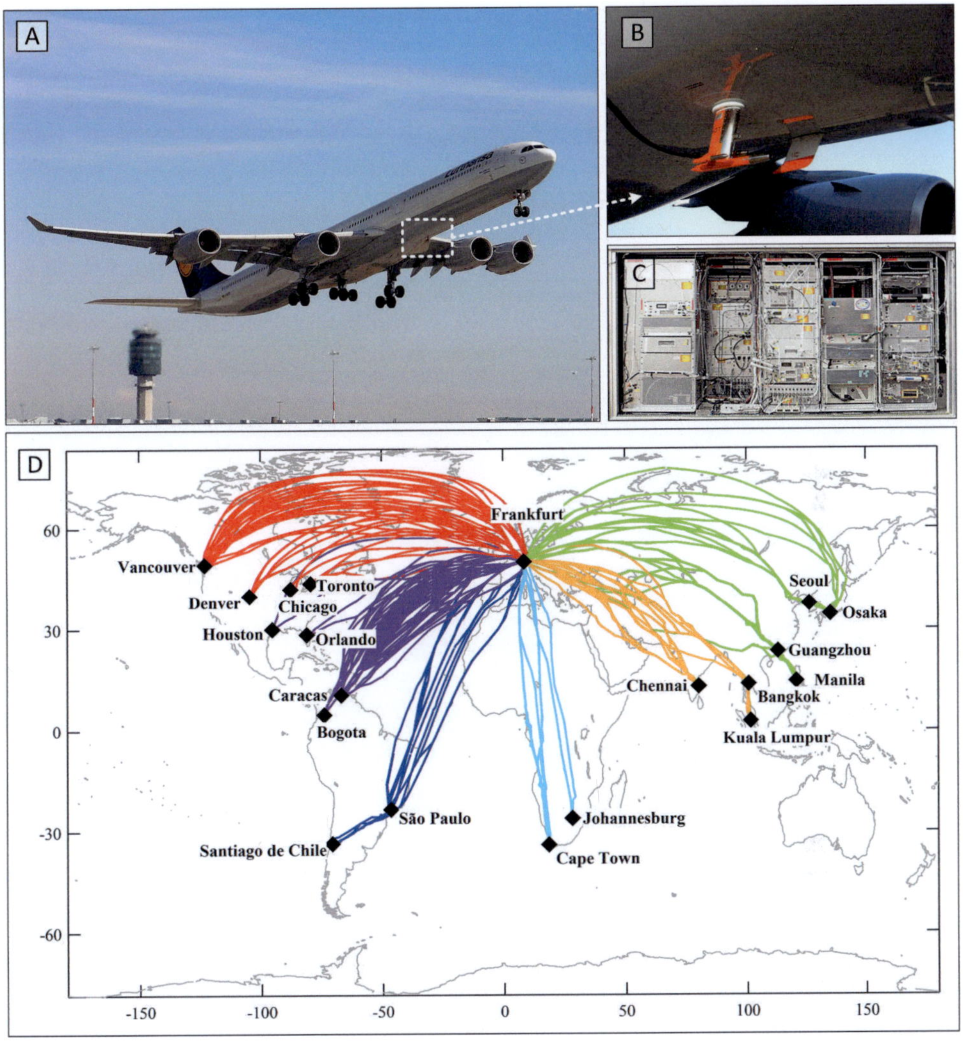

Figure F 5.1: *The CARIBIC project: Panel* A *shows the Lufthansa Airbus A 340-600 'Leverkusen' with mounted trace gas inlet fin (orange-colored) underneath the fuselage in front of the wing section (detailed side-view depiction of the inlet in* B*). Panel* C *shows the front side of the instrument container with the CARIBIC PTR-MS on the left-hand side.The worldwide coverage is illustrated in panel* D *in form of a selection of performed flight routes, that are color coded according different target regions. [photo credits: Lufthansa (panel A); Max Planck Institute for Chemistry (panels B, C, D)]*

5.2 Mission operations aboard the research aircraft HALO

The HALO[33] aircraft (High Altitude and LOng Range) is a strongly modified commercial business jet Gulfstream G550 of the German Science Community [Schumann & Andreae, 2006], that has a maximum range of $\sim 12'500\,km$ and a ceiling cruise altitude of $\sim 15'500\,m$. Its limit of payload amounts to $\sim 2.8\,t$ for scientific instrumentation. These parameters make it one of the preferential European carriers for air quality and climate related measurements especially at higher altitudes. Figure F 5.2 (p. 86) shows the HALO aircraft during a landing approach at its headquarters in Oberpfaffenhofen/Germany (airport EDMO).

Figure F 5.2: *The HALO research aircraft during landing approach at its headquarters in Oberpfaffenhofen/Germany [photo credit: DLR]*

5.2.1 Aeronautical Certification

In order to deploy the ULW-PTR-MS aboard HALO, the aeronautical certification and permission of an official operating license by the German Federal Office of Civil Aeronautics (Luftfahrt Bundesamt - LBA[34]) has been aspired in collaboration with the engineering development company Enviscope[35] (Frankfurt a. Main/Germany)

[33] http://www.halo.dlr.de/
[34] http://www.lba.de/
[35] http://www.enviscope.de/

together with Gomolzig Flugzeug- und Maschinenbau GmbH[36] (Schwelm/ Germany) and comprised mechanical, electrical and safety inspections. In order to attest the system's harmlessness in particular for the electrical/compass aircraft systems, EMI and magnetic field tests have been accomplished. Figure F 5.3 (p. 87) shows the ULW-PTR-MS during radiation testings in an EMI test chamber at the test laboratory Mectronic[37].

Figure F 5.3: *The ULW-PTR-MS (in front view) running in HALO configuration during an EMI test in a RF anechoic chamber with isolator plate and interconnected singleline impedance stabilization network (Type LN-DO160/45 for airborne equipment) (blue) in front of pyramid shaped RAM (radiation absorbent material). [photo credit: Schmidt/Reumann, Mectronic]*

[36] `http://www.gomolzig.de/`
[37] `http://mectronic.de/`

5.2.2 Upcoming Study: Oxidation Mechanism Observations (OMO)

Mission idea The mission Oxidation Mechanism Observations (OMO) is a long-time planned aircraft mission investigating climate-relevant questions concerning free radical chemistry in the atmosphere, which is organized by the department for Atmospheric Chemistry[38] at the Max Planck Institute for Chemistry in Mainz/ Germany. With special attention to HO_x radicals ($OH+HO_2$), convection and inter-continental long-range transport of air pollutants (NMVOC, NO_x), as well as sinks and sources of (greenhouse-active) trace gases and their photochemical processing are of interest. Especially transport (long-distance transport of air pollutants, e.g. Brewer-Dobson/Hadley circulations; episodic convective uplift pollution; upward transfer in the warm conveyor belts of cyclonic disturbances) and oxidation rates of natural and anthropogenic compounds affecting the global distribution of pollutants are to be examined in the extra-tropical free troposphere (ex-FT) up to high altitudes ($\sim 8 - 15\,km$). This is a sensitive region of the atmosphere in terms of climate responses to changes in its chemical composition [Gettelman *et al.*, 2011]. Since the Indian summer monsoon is a prime example of heavy vertical mixing across the entire free troposphere and clearly deep into the lower stratosphere (details about the Asian summer monsoon are given e.g. in Bhaskaran *et al.* [1996]; Cadet [1979]; Scheeren *et al.* [2003]), exactly such scenario has been chosen for these investigations. Additionally, as described by Lelieveld *et al.* [2001], the targeted area over the entire Northern Indian Ocean towards the Intertropical Convergence Zone (at about $6°S$) is disproportionately affected by high pollution levels (due to fossil fuel combustion, biomass burning), that crucially enhances air quality degradation with implications on local, regional and even global scales. Therefore the measurements are considered to take place in Southeast Asia during summer months.

Mission history The mission OMO was initially scheduled to take place in the year 2009. Due to certification and technical completion issues of both the HALO aircraft (i.e. certification delays of the carrier, trace gas inlets, external instrument compartments, etc.) and also scientific instrumentation (i.e. due to unrealistic basic assumptions during the setup phases), OMO has been successively postponed over years already. It is presently assumed that test flights at the HALO headquarters in Oberpfaffenhofen/Germany will take place in January/February 2015 (flight hours: $16\,h$). The destinations for technical flight tests are planned to be above the Baltic Sea (due to short approaching times) and for instrumental flight tests above the Mediterranean Sea (around Malta), where the needed photo-chemistry is enhanced due to the higher solar altitude compared to the Baltic Sea. Likewise, the actual flight planning provides that research flights in the surroundings of the Indian summer monsoon will happen also in 2015 (flight hours: $\sim 80\,h$) originating presumably from Sri Lanka, which would allow measurements in flight corridors West of India towards the United Arab Emirates ($\rightarrow$ monsoon outflow), and East of

[38] `http://www.mpic.de/en/research/atmospheric-chemistry.html`

India towards the Bay of Bengal and Bangladesh (→ relevant area for heavy vertical upward transport). This choice has been made since direct measurements in the Indian airspace will not be authorized.

The ULW-PTR-MS as part of the scientific instrumentation As part of the scientific instrumentation, the special objectives for the ULW-PTR-MS are the measurement of precursors in the HO_x chemistry, like e.g. the photochemically active compound acetone as source of $HO_2^\bullet$ radicals in the upper troposphere and intermediate sink of NO_x via the production of PAN. Likewise, the oxidation of methanol (CH_4O) as source of formaldehyde (CH_2O) also plays a role in the atmospheric HO_x budget since the photochemical decay of formaldehyde produces 2 $HO_2^\bullet$ radicals [Lindinger *et al.*, 1998; Neumaier *et al.*, 2014] (see also section 1.2.2, p. 8). In addition to that, the general investigation of organic air pollutants with particular attention to their altitude distributions is another main focus.

Flight rehearsal and mission preparations Figure F 5.4 (p. 90) shows the instrument ULW-PTR-MS integrated into HALO during the flight rehearsal for OMO in late winter 2013 at the HALO HQ in Oberpfaffenhofen/Germany. Additionally mounted to the same rack are the corresponding HALO power supply (Rack Power Distribution Box - RPDB) in the upper triangle and the ozone device FAIRO [Zahn *et al.*, 2012]. This instrumentation already represents the final setup for the planned measurements during OMO in 2015. It is designated in the cabin layout to be deployed to the first instrument payload bay (right side of aircraft) behind the cockpit in order to make use of the unique features of both small and lightweight instruments as weight counterbalance to other large and heavy equipment allowing a stable attitude of flight. In this context a new aircraft inlet (TGI 320/modification V318) exclusively supplying ULW-PTR-MS/FAIRO with ambient air (providing short inlet distances to the instruments) has been designed and approved beforehand to the mission start. The final instrumental integration time into HALO is scheduled to start in November 2014. Some impressions of the aircraft as received during the flight rehearsal are attached to appendix E (p. 173).

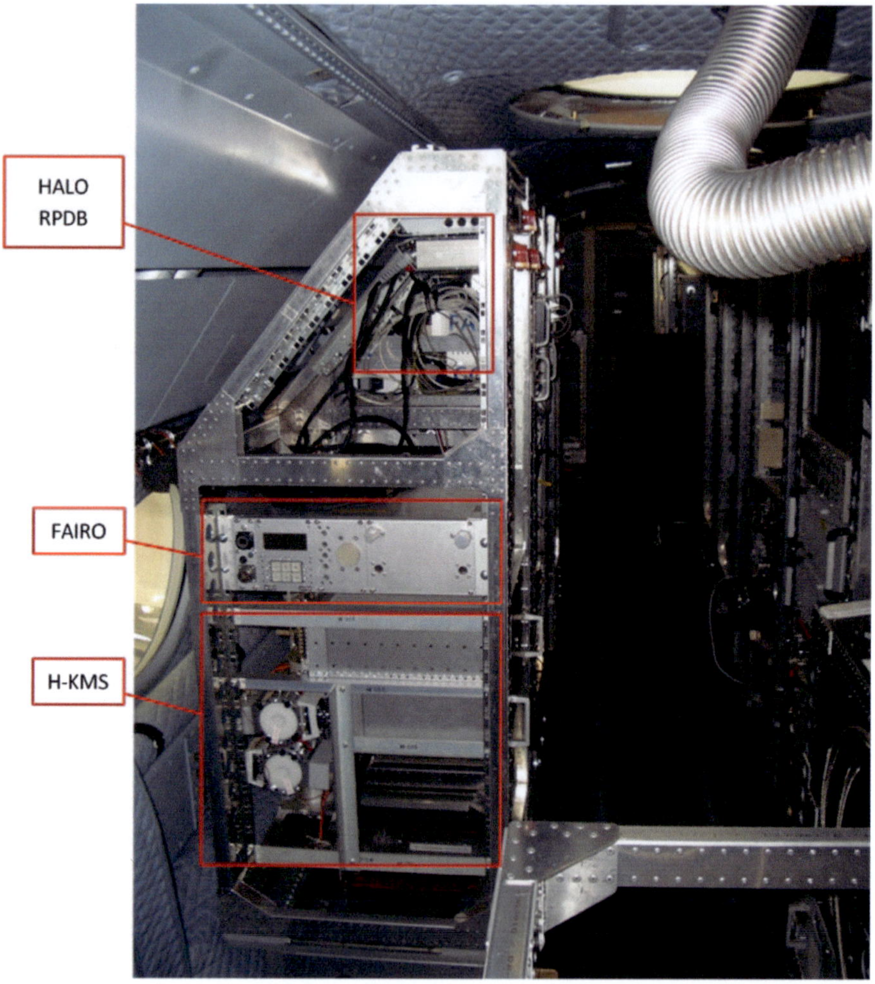

Figure F 5.4: *View along the aisle in the HALO aircraft from the cockpit to the payload bays during the flight rehearsal for the mission OMO in Oberpfaffenhofen/Germany in February 2013. The ULW-PTR-MS (here denoted with its mission-specific identification H-KMS), the ozone device FAIRO and the power distribution device HALO RPDB are mounted in the HALO aircraft rack on the front left.*

Part III

Field Measurements: Air Quality Investigations In The Context Of Fossil Fuel Production

The ULW-PTR-MS was used for the first time during the field mission Energy and Environment - Uintah Basin Winter Ozone Study 2012 (UBWOS 2012), which was an air quality study in the context of oil and natural gas exploitation and production in the Uintah Basin in the rural Northeast of Utah/USA. Comprehensive measurements concerning the atmospheric impact of the oil and natural gas (O&NG) business were performed. The primary task of the study was the investigation of enhanced surface ozone formations and associated precursors during wintertime. Among the multitude of studied parameters, a large data set of VOC emissions was produced while measuring background concentrations and individual emissions at identified point sources in this unusual environment. The results contributed to a better understanding of emission scenarios in connection with O&NG production and in particular to processes affecting the basin's air quality.

This part of the work describes the scientific activities connected with measurements made by ULW-PTR-MS during UBWOS 2012. The chapter 6 (p. 97) gives background information to the O&NG business in the United States and outlines the mission and its objectives. The next chapter 7 (p. 113) presents selected mission results, that comprise general estimations of emissions, stationary measurements at a field site and mobile investigations using a laboratory van in diverse areas of the region affected by the air quality issues to be explained. These examinations were designed to cover the full range of possible emitters, and therein were concentrated onto conspicuous point sources, like e.g. a recently hydraulically fractured gas well. Finally, chapter 7.6 (p. 153) summarizes the presented results and gives an outlook to prospective activities following these measurements.

Chapter 6

Uintah Basin Winter Ozone Study 2012 (UBWOS 2012)

6.1 Background Information

6.1.1 Fossil Fuel Production in the United States

Importance of shales The ongoing improvements in advanced technologies for crude oil and natural gas production, i.e. the use of hydraulic fracturing and horizontal/directional drilling (see section 6.1.4, p. 100), have greatly increased the extraction of fossil fuels within recent years. Especially unconventional reserves in shale formations are of particular importance. The growing relevance of fossil fuels produced from shales is illustrated in figure F 6.1 (p. 98) showing the annual production of dry natural gas by individual source for the US. The data base on historic production records until the year 2011. Future projections are estimated according expected pumping rates of the fossil fuel industry and result in a production of 33.14 trillion cubic feet ($33.14 \times 10^{12}\ ft^3 \simeq 0.94 \times 10^{12}\ m^3$) in total in the year 2040. The latest future estimations by the United States Energy Information Administration[39] (EIA) even forecast a further increase to then 56% in total natural gas production due to increased development of shale gas, tight gas and offshore natural gas resources in the US from 2012 to 2040 to then $37.6 \times 10^{12}\ ft^3$ in total. The significantly highest contributions are expected to come from shale formations. [EIA, 2013, 2014]

Different states - different rules The single US states have individual environmental regulations concerning well operations and the handling of emissions in the context of exploitation and production of fossil fuels. The applied production procedures on site are adapted to the respective legal provisions, and thus differ from state to state and field to field. This particularly means e.g. the mandatory use of vapor capture systems (in form of scrubbing devices) and closed waste water collection containers in Colorado, whereas this was not required in Utah at the time of the presented measurements.

---

39 http://www.eia.gov/

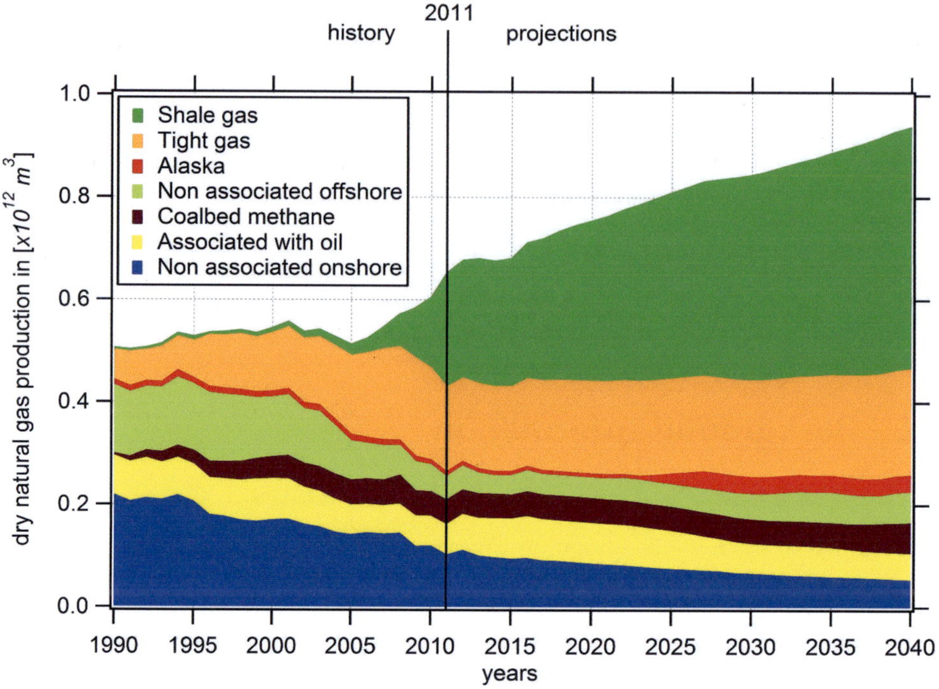

Figure F 6.1: *US dry natural gas production by source in [$\times 10^{12}$ m^3] for the years 1990–2040. (data adapted from U.S. Energy Information Administration - Annual Energy Outlook 2013 Early Release Overview [EIA, 2013])*

Standard emission regulations are provided by the United States Environmental Protection Agency[40] (US EPA) or the respective state programs for fossil fuel administration.

Atmospheric and health impacts of fossil fuel production Intense extraction operations of fossil fuels are generally accompanied by substantial emissions of greenhouse gases, VOCs and nitrogen oxides, which has been demonstrated in several studies, like Gilman *et al.* [2013]; Howarth *et al.* [2011]; Karion *et al.* [2013]; Katzenstein *et al.* [2003]; Kemball-Cook *et al.* [2010]; Litovitz *et al.* [2013]; Pétron *et al.* [2012]. According to the Clean Air Act[41] stated by the US EPA [EPA, 2012], some of the emitted compounds are classified as air toxics that can threaten the health of oil/gas workers and local residents directly [Colborn *et al.*, 2011; McKenzie *et al.*, 2012]. With regard to this work, the general health hazards of the aromatic hydrocarbons benzene (C_6H_6), which is determined to be carcinogenic and mutagenic [ATSDR, 2007],

[40] http://www.epa.gov/
[41] http://www.epa.gov/air/caa/index.html

and toluene (C_7H_8), which is determined to affect fetal development [ATSDR, 2000], are particularly cited as examples.

Regulations and emission limit values as part of protection policies and occupational safety are given in Esswein *et al.* [2014] or can be found in the corresponding databases provided by the National Institute for Occupational Safety and Health (NIOSH[42]) and the United States Department of Labor - Occupational Safety & Health Administration (OSHA[43]), respectively.

6.1.2 The oil and natural gas production in the Uintah Basin, Utah

The Uintah Basin in the Northeast of Utah is an enclosed basin, that is part of the larger area known as Colorado plateau. It is a desertic region with extensive oil and natural gas deposits, that are widely associated with the organic-rich geological Green River formations in the Western United States [Chidsey Jr. & Laine, 1992; Rice *et al.*, 1992; Zhang *et al.*, 2009]. More than 10'000 wells are currently operated in the greater area with ~60 % of them producing natural gas and ~40 % producing crude oil. About 1'000 wells are added each year with proposals in place for tens of thousands more [Helmig *et al.*, 2014; Warneke *et al.*, 2014]. The oil fields are mostly located in Duchesne County and the majority of gas fields can be found in Uintah County.

The Uintah Basin fossil fuel production was characterized by many independent wells without area-wide network of pipelines requiring on road transportation. Methanol, which was on hand at most well pads, was used for dehydration and de-icing during O&NG extraction and to prevent the formation of hydrates. The diverse well pad facilities were primarily driven using diesel or gas generators.

6.1.3 Wintertime observations in the fossil fuel production area - problem definition

During wintertime inversion conditions with snow covering the ground, the measured concentrations of surface ozone in the Uintah Basin frequently reached values clearly above the National Ambient Air Quality Standards[44] (NAAQS) of 75 *ppbV* (averaging over 8 hours). Elevated ambient air O_3 levels were first time reported in the Uintah Basin in the winter 2009/2010 at Ouray, Utah, which is in the center of the basin. Likewise other air toxics have been detected in high mixing ratios as well. Comparable observations have been made in the Upper Green River Valley, Wyoming, close to the Jonah-Pinedale Anticline natural gas field, which is another extensive fossil fuel extraction area with similar topographic and geologic prerequisites. In connection with these observations, Schnell *et al.* [2009] diagnosed

[42] `http://www.cdc.gov/niosh/`
[43] `http://www.osha.gov/`
[44] `http://www.epa.gov/ttn/naaqs/`

a coincidence of enhanced ozone mixing ratios with the production of fossil fuels. Commonly known as customary summertime effect (see section 1.2.1, p. 6), the emitted organic compounds together with nitrogen oxides from diverse O&NG operations enabled ozone formation also in wintertime conditions with enhanced UV intensity due to the high albedo of the snow-covered surface [Carter & Seinfeld, 2012; Pinto, 2009]. A complete overview of the situation and pilot surveys in the basin beforehand to the presented field mission UBWOS 2012 is summarized in the 'Final Report: Uinta Basin Winter Ozone and Air Quality Study / December 2010 - March 2011' [Martin *et al.*, 2011].

6.1.4 Hydraulic Fracturing

Hydraulic fracturing, or short (hydro-)fracking, is nowadays oftentimes used to produce oil and gas in regions where conventional methods, i.e. the opening up of subsurface cavities filled with fossil fuels by direct drilling, is not possible or productive (any more).

The applied technique of controlled horizontal/directional drilling enables the punctual approaching of hydrocarbon rich shales in appropriate schistous rock formations, where gas and oil is assumed to be encased. As the reachable area outgoing from a single borehole can cover huge territories, the separate developing and drilling of new wells can be reduced at the same time.

The controversial discussion in public about the use of this technology and implications for the environment also covers multifaceted economical, political and health aspects. Some argumentations can be found in Asche *et al.* [2012]; Carlton *et al.* [2014]; Colborn *et al.* [2011]; Ellsworth [2013]; Esswein *et al.* [2014]; Jenner & Lamadrid [2013]; Kinnaman [2011]; McIlvaine & James [2010]; McKenzie *et al.* [2012]; Moore *et al.* [2014]; Osborn *et al.* [2011]; O'Sullivan & Paltsev [2012]; Rahm [2011].

Proceeding In order to break the rocks and with that to free the oil or gas, respectively, water under high pressures is pumped down the wellbores. Together with the water the admixed supporting additives sand, cement and chemicals are pressed into the earth. A variety of chemical products (among them methanol, formaldehyde, ethylene glycol, hydrochloric acid, and sodium hydroxide) are used for purposes such as improving fluid viscosity, inhibiting corrosion, and limiting bacterial growth. The fracking fluid flows into the underground formation and fractures it due to the high induced pressures. With that the injected sand and other proppants are carried into the opened fissures and keep them permeable to enable the flowing out of oil and gas. As consequence of the naturally existing pressures in the undergrounds, the used water is afterwards driven back to the surface with portions of released oil, gas, condensate and the injected ingredients of the fluid. This phase typically persists for several days or even weeks. The initial treatment of the back-flowing liquids comprises intermediate storage in closed containers with vapor capture systems or open flow-back ponds, which is dependent on the particular US state regulations. Further processing of the waste water includes the reuse at other hydrofracking sites, an ambient vaporization e.g. in open evaporation ponds, and the final disposal into water injection wells. The flow-back ponds get filled up with soil and leveled out again after the full completion of the wells. A sketch of the process is shown in figure F 6.2 (p. 101).

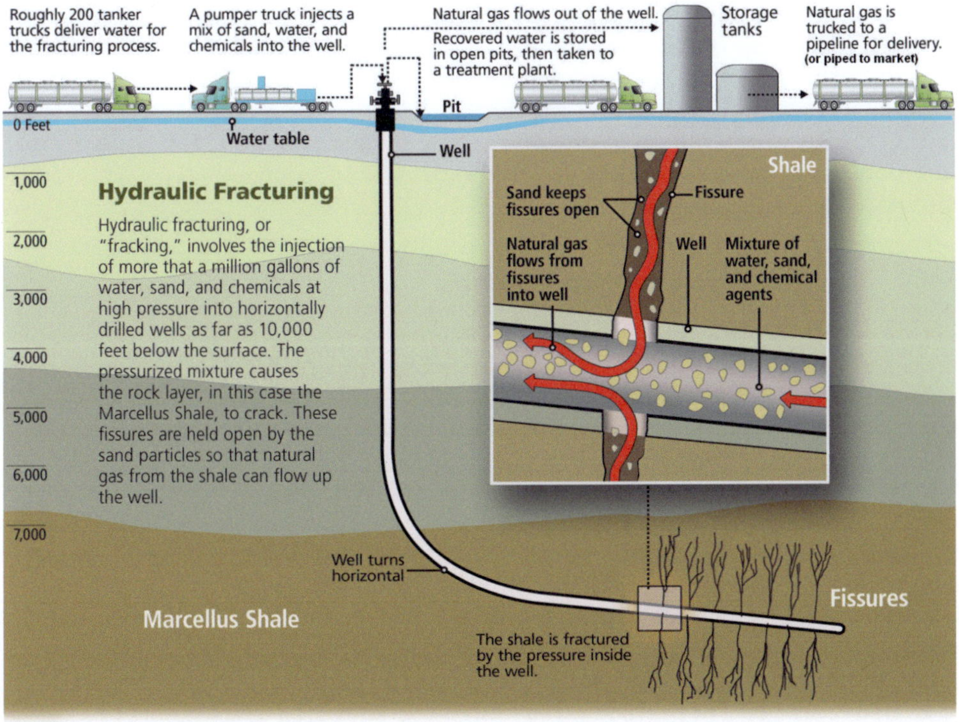

Figure F 6.2: *Overview screen demonstrating a standard hydrofracking process at a natural gas well. (adapted from Manuel [2010])*

The further handling of the final products oil and natural gas then corresponds with standard procedures, comprising the on-site storage in tanks for later transportation or the use of pipelines. Beforehand to that, the well fluids produced from oil and gas wells are normally separated into gaseous and liquid components using pressure vessels. With that, unwanted by-products (e.g. used water, condensate) are segregated from the produced fossil fuels and separately stored. Depending on the moisture of the products themselves and the technical design of the system, an additional dehydration step (e.g. in form of glycol dehydrators) can be implemented.

Used equipment and atmospheric impact In contrast to subsurface processes being investigated in more detail by now, only few knowledge exists about direct atmospheric impacts of hydraulic fracturing. One of the few studies in that respect, a comprehensive investigation of CH_4 emissions published by Allen *et al.* [2013], diagnosed particular emission sources during hydrofracturing of natural gas production sites in the US. The described processes and used equipment are very similar to what was used during the here reported measurements and can therefore be recommended as technical process description.

In another recent publication by Esswein *et al.* [2014], some chemical exposure risks during flow-back operations in unconventional oil and gas extraction are reported. In order to minimize these risks at workplaces and to protect workers, protective mechanisms are discussed.

6.1.5 The flow-back site Glen Bench

Location and identification A single Uintah Basin gas well, that was recently hydraulically re-fractured at that time, was studied in more detail. The investigated well (in the following referred to as Glen Bench site) is located in the Red Wash fields (marked in figure F 6.4, p. 107) at the crossways of Highway 45 and Dead Man Bench/Glen Bench Road ($40.186° N$, $109.329° W$; $1716\,m$ above sea level). In addition to its locally used name RW 31-28B, the well can also be found under its official American Petroleum Institute (API) identification number API 4304715283. Online data (production, history, etc.) are officially available from the 'Utah Division of Oil, Gas and Mining - Department of Natural Resources[45]'.

Well history The well had been originally drilled in the year 1962 and was thereafter temporarily abandoned, while operators changed a few times. The last reactivation took place in the beginning of the year 2012, that was accompanied by a re-drilling and deepening into the Mesa Verde formations. On 9 February 2012 Glen Bench site was hydraulically re-fractured and waste water was produced into the open flow-back pond for subsequent days.

Well layout and accessibility The Glen Bench site was easily accessible from all around during the intensive mission phase for exhaust gas measurements with respect to prevailing wind directions. Pictures of the well with typical fracking equipment are shown in figure F 6.3 (p. 104). The entire scene at that time is illustrated in figure F 7.15 (p. 142) presenting a satellite picture with superimposed measurement

[45] `http://oilgas.ogm.utah.gov/`

data. During the presented measurements the condensate tanks (for storage), the wellhead (borehole), the methanol tank (supply unit for the additive methanol) and the separator (separation device for liquids and gases) were located on a line from West to East. In addition to these standard facilities a temporary condensate tank and separator were placed next to the respective permanent installations in order to handle the initial return flow after the hydraulic fracturing process. The pit for the open flow-back reservoir was excavated in the South of this line-up towards the bypassing dirt road and lined with plastic foil.

(a) *Glen Bench, seen from street*

(b) *Fracking equipment*

Figure F 6.3: *The Uintah Basin gas well* ***Glen Bench*** *in the Red Wash fields during flow-back phase with (a) from left to right: standard condensate tanks, wellhead (in front of truck), methanol tank, separators and flow-back pond (behind earth bank), (b) line-up of some fracking equipment. The pictures were taken on 26 February 2012.*

6.2 Profile of the campaign

Mission objectives The objective of UBWOS 2012 was the investigation of local ozone chemistry around fossil fuel production sites during wintertime. General air quality measurements were performed with special attention to precursors in the O_3 formation chemistry (see section 1.2.1, p. 6). Therefore particular sources of VOCs, NO_x and particles, unique radical sources, like e.g. nitrous acid (HONO) and formaldehyde (CH_2O), vertical structures and the role of dynamics were in focus during the investigations.

Involved partners and used infrastructures UBWOS 2012 was organized by the Chemical Science Division (CSD) at the Earth System Research Laboratory (ESRL) at the National Oceanic and Atmospheric Administration (NOAA) in Boulder, Colorado/USA and included partners from research and regulatory agencies, scientific institutions and cooperating industry partners (among others: Western Energy Alliance[46]). Further references concerning participants, instrumentation, infrastructures and activities connected to the study can be found at the mission home at NOAA ESRL (UBWOS 2012[47]).

The study included the ground site Horse Pool ($40.143°N, 109.468°W$; $1530\,m$ above sea level), which was fully equipped with a large suite of gas phase and aerosol measurements. It was set up on the Northern slopes of the most extensive Uintah Basin gas fields and instruments were operated during an intensive measurement phase of ~ 6 weeks (15 January - 1 March 2012). The field site was in the main focus for the broad-based investigations and was used as reference station for mobile measurements performed in the various areas of the basin. Longer term ozone probings using tethered sondes and a network of monitor stations, as well as other trace gas data loggings using filter cartridges had been performed in advance by local agencies.

Besides the stationary measurements at Horse Pool, an equipped Mobile Laboratory initiated by the Global Monitoring Division (GMD) at NOAA (further referred to as mobile laboratory) performed several weeks of measurement drives across the basin in order to investigate trace gas distributions and emissions of point sources, like e.g. the recently hydraulically fractured gas well Glen Bench site. Likewise, an instrumented single-engine Mooney TLS aircraft was used to investigate longer-range dynamics and transport of single emission proxies, like CH_4.

Study area Figure F 6.4 (p. 107) shows the Uintah Basin with particular attention to the production of fossil fuels. Oil and gas wells together with other point sources are indicated by color coded markers. The Horse Pool field site at the Northern edge of the Uintah County gas fields and the Glen Bench flow-back site in the Red Wash

[46] http://www.westernenergyalliance.org/
[47] http://esrl.noaa.gov/csd/groups/csd7/measurements/2012ubwos/

fields, that were investigated in detail, are marked. Additionally, major drive tracks of the mobile laboratory are shown in red to indicate the coverage of this study. The drives were concentrated to the gas fields in the South of Horse Pool, but for comparison reasons were conducted also to oil fields in Duchesne County (Western route) and to gas fields in Rangely, Colorado (Eastern route). It has to be mentioned that the depicted fields were not exclusively consisting of either oil or gas wells, but have been predominantly composed of the particular production type.

Weather situation - implications for the investigations in 2012 The winter in that year 2012 was unseasonably warm without continuous snow-cover and expected cold-pool inversion events in the basin [Martin *et al.*, 2011]. The boundary layer was mainly mixed because of daytime surface heatings that broke up the night-time surface inversions. These conditions effectively diluted the observed accumulated surface emissions [Helmig *et al.*, 2014]. Since the weather or snow conditions, respectively, did not provide the expected enhanced photochemistry during the intensive phase in 2012, smaller successor missions (with reduced instrumentation) took place in the following years with particular focus onto this objective (UBWOS 2013[48]; UBWOS 2014[49]). Nevertheless, the conditions in 2012 were especially amenable to study VOC emissions with the mobile laboratory. Extensive measurement drives in different regions inside of the basin were performed in order to investigate the contributions of single point source classes (e.g. oil wells, natural gas wells, storage tanks, compressor stations, processing plants, drilling sites, evaporation ponds, flow-back sites). Results of these mobile VOC surveys enabled the identification of particular emission sources. Moreover, the comprehensive set of observations in 2012 allowed the testing of the O_3 photochemistry understanding in this unusual emissions environment by simulating respective emission scenarios.

The range of ambient conditions during the measurement phase comprised changes in pressures from $P_{min} = 825.85\,hPa$ to $P_{max} = 856.99\,hPa$, relative humidities from $RH_{min} = 10.3\,\%$ to $RH_{max} = 100.2\,\%$ and temperatures from $T_{min} = -14.5\,°C$ to $T_{max} = 13.0\,°C$ ($1\,min$-averages; taken at the Horse Pool walk-up tower $13\,m$ above ground; see figure F 6.5, p. 110).

Some impressions of the region, current conditions and operations at the time (as received during UBWOS 2012) are attached to appendix F (p. 175).

48 http://esrl.noaa.gov/csd/groups/csd7/measurements/2013ubwos/
49 http://esrl.noaa.gov/csd/groups/csd7/measurements/2014ubwos/

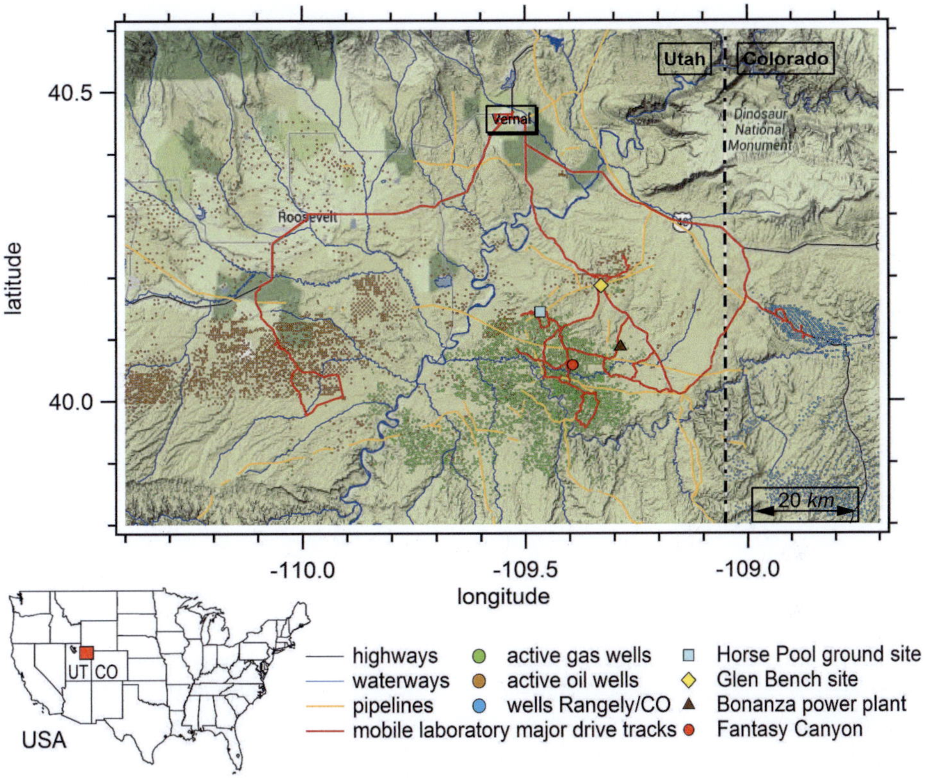

Figure F 6.4: *The map of the Uintah Basin study area shows the locations of oil and natural gas wells, the Horse Pool ground site and the flow-back site Glen Bench. Major routes of the drives with the mobile laboratory are indicated by red lines.*

6.3 ULW-PTR-MS operations

6.3.1 Deployment and setups

The ULW-PTR-MS was deployed at the ground site Horse Pool and inside the mobile laboratory for several weeks each during the intensive measurement phase of UBWOS 2012. In the following only the systems/instruments used for this study are described in more detail.

1. **Stationary measurements: Horse Pool ground site**
   (14 January - 14 February 2012)

   At the Horse Pool ground site (see figure F 6.5, p. 110) the ULW-PTR-MS was deployed in parallel to the other VOC instruments NOAA PTR-MS [de Gouw & Warneke, 2007] and NOAA GC-MS [Goldan, 2004] (discussed in detail in chapter 4, p. 57). The instruments were installed inside a heated laboratory trailer (in the picture: behind the scaffold). The ULW-PTR-MS shared an inlet line with the NOAA GC-MS, which allowed immediate comparisons. The ambient sampling probes were mounted to the erected open scaffold in order to avoid interferences with other infrastructures on site. This additionally enabled a change of inlet heights along the structures. The inlet lines were connected to reversed funnels as inlet pipes to prevent from water and dust and were heated in order to suppress condensation and memory effects in the lines.

2. **Mobile measurements: Mobile laboratory**
   (14 - 28 February 2012)

   Halfway through the mission the ULW-PTR-MS was moved to the mobile laboratory for open road measurements. An installation was possible due to its compactness and started initially as experiment. The mobile laboratory is shown in figure F 6.6 (p. 111). It was equipped with standard instrumentation for the meteorological parameters wind speed/direction (2D ultrasonic anemometer), temperature and humidity, and with GPS tracking systems. A commercially available cavity ringdown instrument was used to measure CH_4, CO, CO_2 and water vapor (standard 4-channel Picarro instrument - model 2301-f flight analyzer [Crosson, 2008]). The two-channel cavity ring-down spectrometer instrument NOxCARD (Nitrogen Oxide Cavity Ring Down; described by Fuchs *et al.* [2009]) was deployed for the direct measurement of NO_2 by optical extinction at $405\,nm$ and of either NO_x (= NO_2 + NO) or O_x (= NO_2 + O_3) after chemical conversion to NO_2 (measurement frequency $1Hz$). Standard O_3 monitors using UV absorption at $254\,nm$ (2B Technologies; measurement frequency $0.1\,Hz$) and whole air samplers (total air compositions for subsequent laboratory analysis) were in use. The entire instrumentation was powered by independent battery power supply, which facilitated stationary measurements without interferences due to own car exhausts. A diagonally upwards towering cantilever arm, that was installed to the vans' roof structures, allowed the installation of the inlet lines ahead of the vehicle body. The probes (covered by dust protection) had a sampling height of $\sim 4\,m$ above ground and were connected to the instruments in the inside using PFA tubes. In order to allow only short inlet time delays, an additional sucking pump acting as bypass line was installed.

6.3.2 Operating characteristics

ULW-PTR-MS settings The instrumental settings used during the mission were a drift tube pressure $P_{drift} \simeq 2.2\,hPa$, a drift voltage $U_{drift} = 612\,V$ and water vapor flows $F_{H_2O} = 7.5\,sccm$ into the IS. The used instrumental configuration at that time resulted in an average primary ion yield of $3 - 5 \times 10^6\,cps$ and a general sensitivity of around $100 - 200\,cps\cdot ppbV^{-1}$. The comprehensive characterization is presented in detail in section 4 (p. 57).

Measurement sequences and system calibrations The ULW-PTR-MS was operated in two modes: ① full scanning from m/z 20 to m/z 169 with $1\,sec$ per m/z (cycle time: $150\,sec$), ② 37 selected m/z (see table T 4.2, p. 64) at first measured for $\sim 1sec$ per m/z (cycle time: $39\,sec$) and later for $0.5\,sec$ every $\sim 19\,sec$. The faster duty cycle was introduced in order to reach better spatial resolution during the mobile laboratory drives. Internal background calibrations were typically performed every $30\,min$ and external bottle calibrations at the Horse Pool field site every other day (see section 4.1, p. 57).

Performance aboard the mobile laboratory In spite of the harsh conditions for operations in the mobile laboratory, hardly any significant deteriorations of accuracy were determined besides a negligible increase of instrumental noise. Nevertheless, the use aboard this moving platform driving on dirt roads in difficult terrain complicated some instrumental operations in consequence of ever-changing operating conditions, i.e. vibrations and omnidirectional accelerations to the instrument. For this reason single sequences were excluded from the further analysis due to anomalies in the house-keeping data. For example one could mention fluctuations in the TMP rotation speeds (typical: $\sim 90\text{'}000\,rpm$) as consequence of heavy shocks to the instrument and with that a possibly varying P_{drift}. However, as the presented measurements at point sources were exclusively performed in slow walking speeds, the presented data are expected to be free of interferences.

Figure F 6.5: *The field site Horse Pool with major part of the stationary measuring instrumentation distributed to several instrument containers, the open scaffolding and other outside equipment (like e.g. radiometer tower). The picture was taken on 4 February 2012.*

Figure F 6.6: *The mobile laboratory in front of the Glen Bench site during sudden fall in temperature on 19 February 2012. [photo credit: Gabrielle Pétron, NOAA]*

Chapter 7

Results and Discussion: Characterization of VOC emissions & compositions in the Uintah Basin

In this chapter the VOC measurements during the mission UBWOS 2012 are presented and analyzed. The first section 7.1 (p. 113) reports about some general results for placing the data into a broader context. The section 7.2 (p. 117) describes stationary measurements at the Horse Pool field site. In section 7.3 (p. 122) a comparison of VOC spectra is shown. The following sections present results from measurement drives with the mobile laboratory, that comprise investigations of emissions from diverse point sources (section 7.4, p. 126), comprehensive investigations of emissions at the flow-back site Glen Bench (section 7.5, p. 132) and the hydrogen sulfide distribution in the basin (section 7.6, p. 149).

7.1 General results - introductory remarks

7.1.1 The VOCs' functions in the basin's chemistry

Chemically reactive environment The broad suite of observed pollutants at partially very high mixing ratios made the Uintah Basin to a chemically reactive environment. Box model simulations by Edwards *et al.* [2013, 2014], that were each constrained to the observations during the missions UBWOS 2012 and UBWOS 2013, identified the O_3 production photochemistry to be highly conditioned by the individual roles of VOCs and NO_x and with that directly dependent on the sources of radicals controlling the oxidation cycles (see also section 1.2.1, p. 6). As further discussed in both publications, especially radical amplification reactions during the oxidation of diverse VOCs contributed to this sensitiveness of the ozone formation chemistry in this unusual environment. The chemical mechanisms governing the (wintertime) O_3 formation for the specific case of the Uintah Basin were therefore determined to be directly connected to the emissions of organic trace compounds. Thus, the identification of emission sources of VOCs (and NO_x) from O&NG production was key to understanding the build-up processes in order to explain the causes leading to the enhanced wintertime ozone formation.

The role of snow cover The investigations of the two central questions concerning the determination of chemical mechanisms under appropriate conditions, i.e. with or without cold pool events, and a direct identification of individual emission sources have been governed by the different weather situations in the successive winters of 2011/2012 and 2012/2013 (see section 6.2, p. 105). Whereas the first winter was mostly snow-free (particularly enabling extensive mobile measurements as presented in the following sections), multiple strong O_3 events occurred only in the colder and consistently snow covered winter in the following year. Thus, the question of the role of snow cover in driving the O_3 formation raised, as its principal effects are to increase surface albedo and with that actinic flux for photolysis reactions. Likewise, the reduction of the mixed-layer height, which concentrated primary emissions, played also a role.

The comparison of the box model simulations of both years showed, that comparable VOC mixing ratios (same estimated high VOC emissions for both years; higher dilution in 2012) with different albedo leads to dissimilar O_3 formation with clearly higher rates in the second winter. This demonstrated that both high VOC emissions into the shallow and stable boundary layer together with increased photolysis rates due to the snow albedo are necessary for the observed rapid winter ozone production. [Edwards *et al.*, 2013, 2014]

Carbonyl photolysis as driving radical source In continuation to the box model simulations of the snow-free season, which confirmed the O_3 production photochemistry to be highly radical limited and the system to be VOC sensitive, the comparable simulation study constrained to the observations of the second season finally determined the winter ozone pollution to be driven primarily by photolysis of carbonyl compounds. Edwards *et al.* [2014] further reported that the extreme VOC concentrations in this unique environment optimized the ozone production efficiency of NO_x (at lower NO_x and much larger VOC mixing ratios than in its summertime urban counterpart). This led the carbonyl photolysis being a dominant oxidant source producing significant numbers of radicals in addition to the primary radical sources (as shown in section 1.2.2, p. 8). The corresponding reaction channel of the simplest example of this process is reported to be the high-NO_x oxidation of methane, which starts with an OH radical attacking the CH_4 in the presence of O_2 to produce a peroxy radical and water:

$$OH + CH_4 + O_2 \longrightarrow CH_3O_2^\bullet + H_2O \qquad \text{(R7.1)}$$

The peroxy radical then reacts with NO to an alkoxy radical:

$$CH_3O_2^\bullet + NO \longrightarrow CH_3O^\bullet + NO_2 \qquad \text{(R7.2)}$$

The following reaction of the alkoxy radical with O_2 then leads to the formation of formaldehyde (HCHO) and a hydro-peroxy radical:

$$CH_3O^\bullet + O_2 \longrightarrow HCHO + HO_2 \qquad \text{(R7.3)}$$

Finally the hydro-peroxy radical reforms an OH in reaction R7.5 and the HCHO, as chemically stable product of the radical propagation chain, photolyses readily to 2 hydro-peroxy radicals via its dominant channel (yield $\sim 70\,\%$ according Carbajo *et al.* [2008]):

$$0.7\times(\mathrm{HCHO} + h\nu + 2\,\mathrm{O_2} \longrightarrow \mathrm{HO_2} + \mathrm{HO_2} + \mathrm{CO}) \qquad \text{(R7.4)}$$

In addition to the radical that is propagated through this mechanism, the primary radical source is amplified by means of formation of 2 additional radicals:

$$2\times(\mathrm{HO_2} + \mathrm{NO} \longrightarrow \mathrm{OH} + \mathrm{NO_2}) \qquad \text{(R7.5)}$$

It was determined that different species constituted the carbonyl radical source. Formaldehyde, which is due to the oxidation of all VOCs (in particular from light alkanes), was identified to be the single largest contributor to this source. Likewise, aromatic compounds were also found to be an important origin for radicals. [Edwards *et al.*, 2014]

The chemical cycle describing the ozone formation is presented in detail in section 1.2.1 (p. 6).

7.1.2 Estimations of methane and VOC emissions

The total emissions originating from the Uintah Basin were estimated using data from light aircraft methane measurements and balloon-borne vertical in situ VOC profile measurements. Two studies, that were performed in parallel to the ground-based activities, are quoted in order to give a brief review of the dimensions of the emissions:

- **CH_4:** Based on a mass balance approach using airborne measurement data, that were obtained during a research flight on 3 February 2012, Karion *et al.* [2013] estimated an average methane leakage rate of $(8.9 \pm 2.7)\,\%$ of production with total CH_4 emissions of $(54.6 \pm 15.5) \times 10^3\,kg\cdot h^{-1}$ for the Uintah Basin.
- **VOCs:** An estimation of organic trace gas emissions for the Uintah Basin using tethered balloon measurements was performed by Helmig *et al.* [2014]. The reference period was between January - February of 2012 and 2013 and total annual mass fluxes of C_2 - C_7 VOCs into the basin with $(194 \pm 56) \times 10^6\,kg\cdot yr^{-1}$ were reported. This is equivalent to the annual emissions of a standard fleet of ~ 100 million cars. In this context the respective total annual fugitive emissions of the aromatic hydrocarbons benzene and toluene into the basin were estimated to be $(1.6 \pm 0.4) \times 10^6\,kg\cdot yr^{-1}$ and $(2.0 \pm 0.5) \times 10^6\,kg\cdot yr^{-1}$, respectively.

7.1.3 Comparison with typical urban air compositions

VOCs observed in the Uintah Basin clearly differed in composition and distinctly exceeded in mixing ratios with typical urban air compositions, as e.g. reported by Mohamed *et al.* [2002] (comparison of 13 US cities), de Gouw [2005] (ship-based air quality measurements in the New England region), Borbon *et al.* [2013] (anthropogenic emissions in the Northern mid-latitude megacities Paris and Los Angeles), Karl *et al.* [2009] (New Mexico) or Warneke *et al.* [2013] (temporal weekday-weekend effects in the Los Angeles basin). The occasionally measured peak values in the Uintah Basin (presented in section 7.5.1.2, p. 134) outnumbered typical urban average concentrations for aromatic hydrocarbons by several orders of magnitude (factors of $\sim 10^2 - 10^3$).

Further portraits of organic trace gas compositions originating from O&NG operations have been presented by Gilman *et al.* [2013] (source signatures in Northeastern Colorado) and Yuan *et al.* [2014] (airborne measurements downwind of the Deepwater Horizon oil spill in the Gulf of Mexico). Reported compositions and the range of typical mixing ratios mainly conform with findings from measurements in the Uintah Basin.

7.2 Typical VOC mixing ratios and compositions at the field site Horse Pool

The measuring site Horse Pool, though being located in the fringe zone of the basin's main production area (see map in figure F 6.4, p. 107), was surrounded by ~ 160 individual well pads within a $5\,km$ radius [Helmig *et al.*, 2014]. Both background concentrations and short-term spikes of emissions from neighboring trace gas point sources were detected.

Longterm series Full time series of selected organic trace compounds (benzene (m/z 79), toluene (m/z 93), methylcyclohexane (m/z 97), dimethylcyclohexane (m/z 69, m/z 111), methanol (m/z 33)) from measurements with the NOAA PTR-MS as presented in Warneke *et al.* [2014] (averaged to the mission timebase of $1\,min$ data) are shown in figure F 7.1 (p. 119). Methane mixing ratios as prime proxy for natural gas emissions are included to illustrate correlated occurrences. The observed mixing ratios were characterized by high values for longer time periods when the air exchange stagnated in the basin. This led to an accumulation of the emissions and was essentially governed by local climate, weather and season. Very high mixing ratios during short-term spikes were found, that were individually caused by local point sources. These spikes were typically related to processes in connection with close O&NG production activities or associated with therewith connected heavy traffic on the bypassing access route. Frequent toluene spikes clearly above $10\,ppbV$ were found. Benzene and toluene mixing ratios of over $3\,ppbV$ from 30 January to 2 February 2012 were observed, before the basin cleaned out and concentrations dropped to typical continental background values during wintertime. Methanol was almost an order of magnitude higher than the detected aromatic hydrocarbons with spikes even reaching up to several hundreds of $ppbV$.

Air mixtures measured at Horse Pool Horse Pool was predominantly surrounded by gas wells with some oil wells in the Northwest. The winds mainly comprised Southern components, which is demonstrated in form of a rose plot (with $5°$ sectors) in figure F 7.2 (p. 120) showing the directional distribution of benzene mixing ratios (used as proxy for pollution). The data represent the entire mission phase and distances to the origin indicate relative probabilities per $5°$ sector. No convolution according to wind speeds was used here, as only the directional distribution and relative differences of the fractions are to be shown exemplarly. When comparing the directional distributions of various compounds measured at the field site, two different compositions of air were found and enabled a segregation of the data into two classes of emissions. A small fraction of $\sim 5\%$ was different to the rest showing enhanced C_4 - C_{11} n-alkanes relative to propane with significantly lower ethane to propane ratios as measured by NOAA GC-MS [Warneke *et al.*, 2014]. These air mixtures were almost exclusively associated with air coming from Northwestern directions, where two oil well pads were located nearby (referred to as oil well sector). The main part of the data of $\sim 95\,\%$ originated from all other directions, where the majority of the Uintah Basin natural gas wells were located (referred to as gas well sector).

Correlation plots of selected VOCs versus benzene (averaged $5\,min$ data), as presented in figure F7.3 (p. 121), were used to compare the measured air mixtures by means of their unique chemical fingerprints. The data from Horse Pool showed good correlations for all compounds with benzene ($R^2 > 0.91$), especially for the gas well sector. In the oil well sector the correlation slopes of the larger aromatics and the cycloalkanes with benzene showed clear differences to the gas well sector. These differences increased with the molecular weight of the compounds. To verify the assumptions of two different source sectors, slopes of measurements obtained with the mobile laboratory at individual point sources, i.e. downwind of oil and natural gas well pads (see section 7.4, p. 126), are added in this illustration. The contrasting of Horse Pool data and identified point source emissions showed comparable slopes and differences between oil and gas wells. But as the measured individual well pads had larger differences among each other, also their variabilities were larger. The comparison of enhancement ratios (ER) over benzene from both emission groups, i.e. the slopes of the linear fits, showed that oil well pad emissions were characterized by clearly larger ERs than the gas well pad emissions and their differences increased with molecular weight. This showed that VOC mixtures emitted from oil wells consisted of heavier compounds than the VOC mixtures emitted from gas wells [Warneke *et al.*, 2014].

Although it was possible to differentiate the air mixtures measured at the field site in order to grossly relate compositions to emission sources, a clear identification of the sources themselves was necessary to estimate individual contributions to the measured compositions.

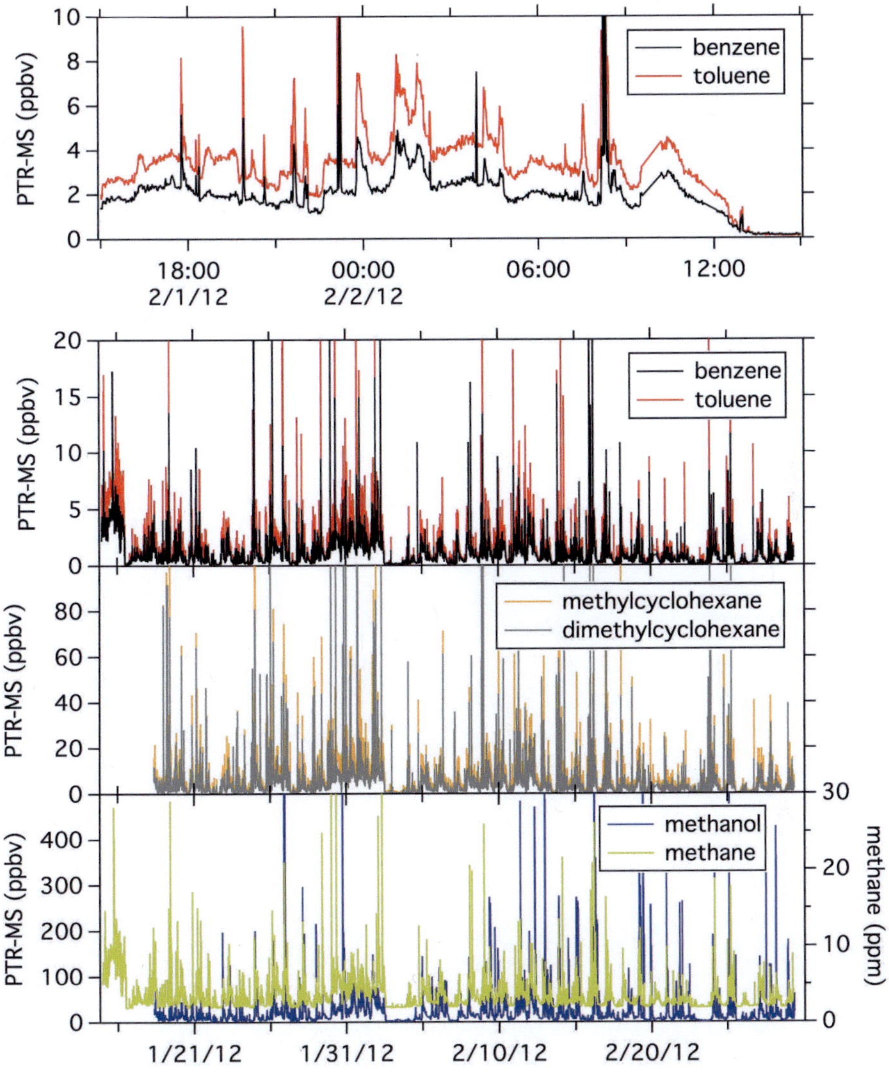

Figure F 7.1: *The time series of various VOCs measured by PTR-MS and methane at the Horse Pool ground site for the complete measurement period during the UB-WOS 2012 campaign. The top panel shows a 24 hour period of benzene and toluene measurements. Maximum mixing ratios are off-scale. (adopted from Warneke* et al. *[2014])*

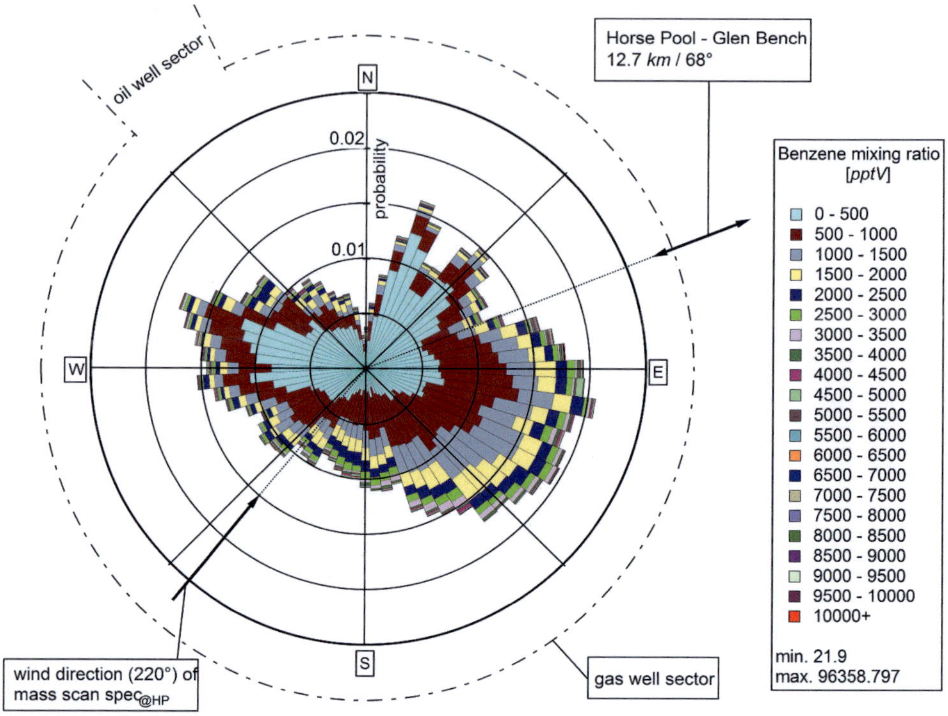

Figure F 7.2: *The rose plot shows the benzene mixing ratios in [pptV] per* 5° *sector measured at the Horse Pool field site with the NOAA PTR-MS. The data represent the entire mission phase and the axis show probabilities per sector. The mass scan* $spec_{@HP}$ *is further discussed in section 7.3 (p. 122).*

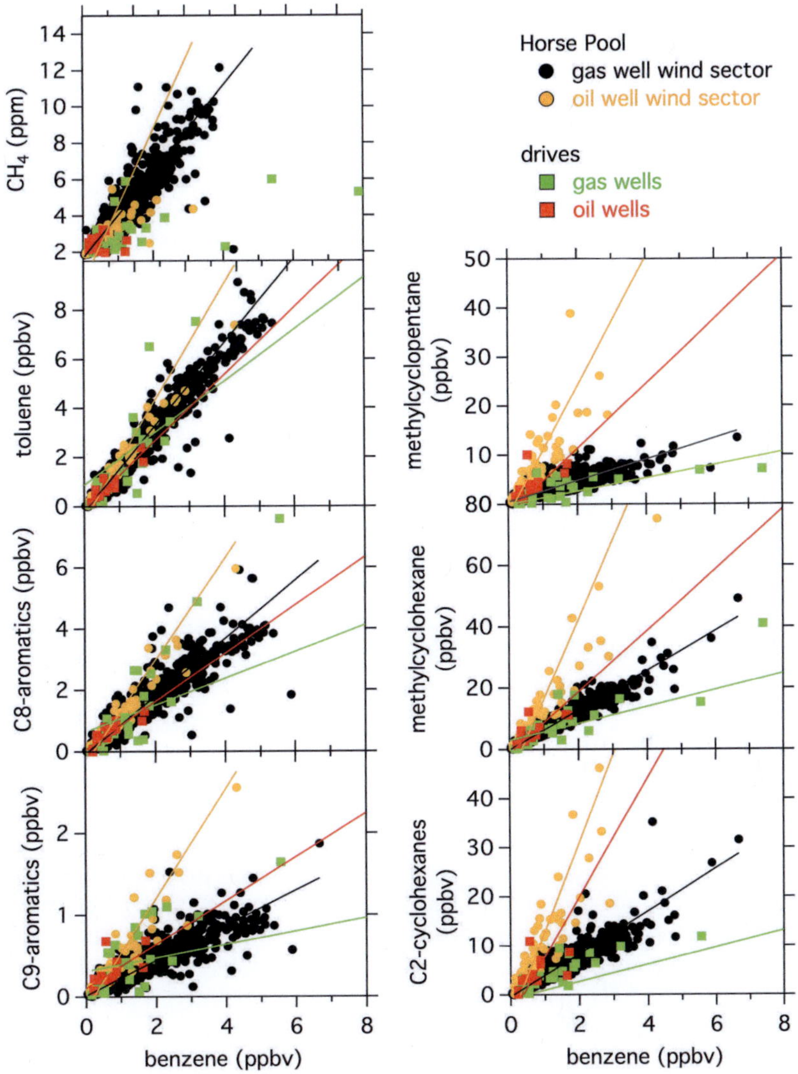

Figure F 7.3: *Scatter plots of toluene, C_8- and C_9-aromatics, methylcyclopentane, methylcyclohexane and C_2-cyclohexanes versus benzene in [ppbV] and methane versus benzene in [ppm] for measurements at the Horse Pool ground site (gas well wind sector in black; oil well wind sector in orange) and mobile laboratory measurements close to individual natural gas well pads (green) and oil well pads (red). (adopted from Warneke* et al. *[2014])*

7.3 Comparison of typical VOC compositions

The VOC composition in the basin was composed from accumulated emissions from the various point sources. In order to derive unique fingerprints of air mixtures measured at peculiar locations, mass spectra from m/z 20 to m/z 169 were recorded during periods with stable winds. This allowed conclusions about the homogeneity of the air mixtures sampled in the basin. The analysis was performed using raw data in $[cps]$, as otherwise all compounds, that were not explicitly calibrated, should have been excluded from the comparisons. However, these comparisons only allow the investigation of relative differences between scans. Moreover, in order to account for the different framework conditions of the used measurements, the axis of the shown data were scaled to benzene signals of $100\,cps$ for an easier comparability.

Fingerprints - Horse Pool vs. Glen Bench vs. South Louisiana light crude The mass spectra shown in figure F7.4 (p. 123) were obtained at the Horse Pool field site (spec$_{@HP}$; 30 January 2012 07:37:04 MST) and at the Glen Bench site (spec$_{@GB}$; 18 February 2012 18:10:51 MST). The presented spec$_{@HP}$ originated from air coming from Westerly directions ($\sim 220°$), where mostly gas wells with single oil wells in between were located (see also figure F7.2, p. 120). The wind speed was around $\sim 0.4\,m \cdot s^{-1}$ during the scan. Spec$_{@GB}$ was obtained downwind of the well area with a wind field comprising a superposed plume from condensate tanks and flow-back pond. In addition to the two Uintah Basin spectra, a reference sample spectrum (spec$_{ref}$) was used, that was taken from the headspace of crude oil during laboratory measurements with the NOAA PTR-MS (further described in Yuan *et al.* [2014]). The sample oil was light crude oil from South Louisiana/USA, which was similar to the oil found in the Uintah Basin. The chemical groups of VOCs (alkanes, aromatics, cycloalkanes and oxygenates) are color coded in the scans.

At Horse Pool high signals of alkanes (e.g. m/z 57, 71, 85), aromatics (m/z 79, 93, 107), cycloalkanes (e.g. m/z 69, 97, 111) and methanol (m/z 33) were observed. The Glen Bench scan showed, that most found organic compounds at the flow-back site were also alkanes (e.g. m/z 71, 85), cycloalkanes (e.g. m/z 97, 111) and aromatics (e.g. m/z 79, 93, 107, 121). The highest signals of oxygenated VOCs during this scanning came from methanol (m/z 33).

In addition to observed distinct occurrences of aromatic compounds (m/z 79, 93, 107, 121, 135, 149, 163) and oxygenated masses (m/z 33, 45, 59, 61, 73, etc.), the scans showed particular abundance of compounds with a mass-to-charge ratio of (m/z) $14 \times n \pm 1$ ($3 \leq n \leq 8$) (i.e. m/z 43, 57, 69, 71, 83, 85, 97, 99, 111). This was in general accordance with other measurements of fossil fuel vapors as reported by Yuan *et al.* [2014], and therewith stood in clear contrast to typically observed spectra from urban areas.

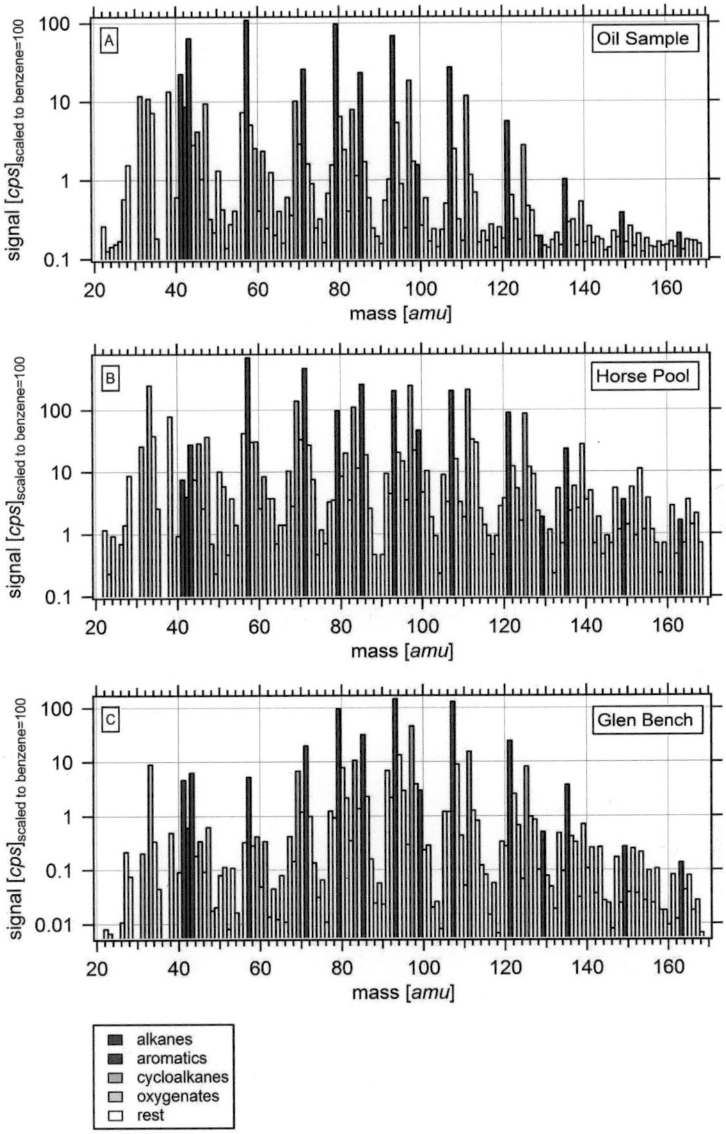

Figure F 7.4: *The mass spectra (m/z 20 - m/z 169 in [amu]) obtained* A *from a South Louisiana light crude [Yuan et al., 2014],* B *at Horse Pool (30 January 2012 07:37:04 MST) and* C *at Glen Bench (18 February 2012 18:10:51 MST) are color coded according to the VOC classes alkanes, aromatics, cycloalkanes, oxygenates. All signals are given in [cps] and were scaled to* $i(benzene) \equiv 100\,cps$ *for a better comparability.*

Correlations of the individual fingerprints Figure F7.5 (p. 125) compares the PTR-MS mass spectra measured at Horse Pool and Glen Bench with that from the crude oil sample using scatterplots, that show the mass spectra versus one another with each symbol representing one mass. It should be considered, that the 1-to-1 line is only inserted to simplify the comparisons between logarithmic and linear projections and does not have any physical meaning. Differences between the scans are recognizable in form of masses, that clearly deviate from the majority of the other masses. The individual VOC fingerprints showed similar characteristics, which was obvious as they consisted of evaporated constituents of fossil fuels. However, although being connatural in the sense of their origins, no reproduction of either of the scans among themselves was observable. This indicated that the VOC mixtures of the vapors at the probed locations and the crude sample, respectively, were heterogeneous.

Despite the demonstrated differences within the mass spectra, the general results of the VOC composition comparisons suggested that the measured air at Glen Bench and at Horse Pool were dominated by vapors from crude oil, but intermixed with specific portions of VOC classes, like e.g. the intense appearances of aromatic compounds at Glen Bench or typical alkanes at Horse Pool.

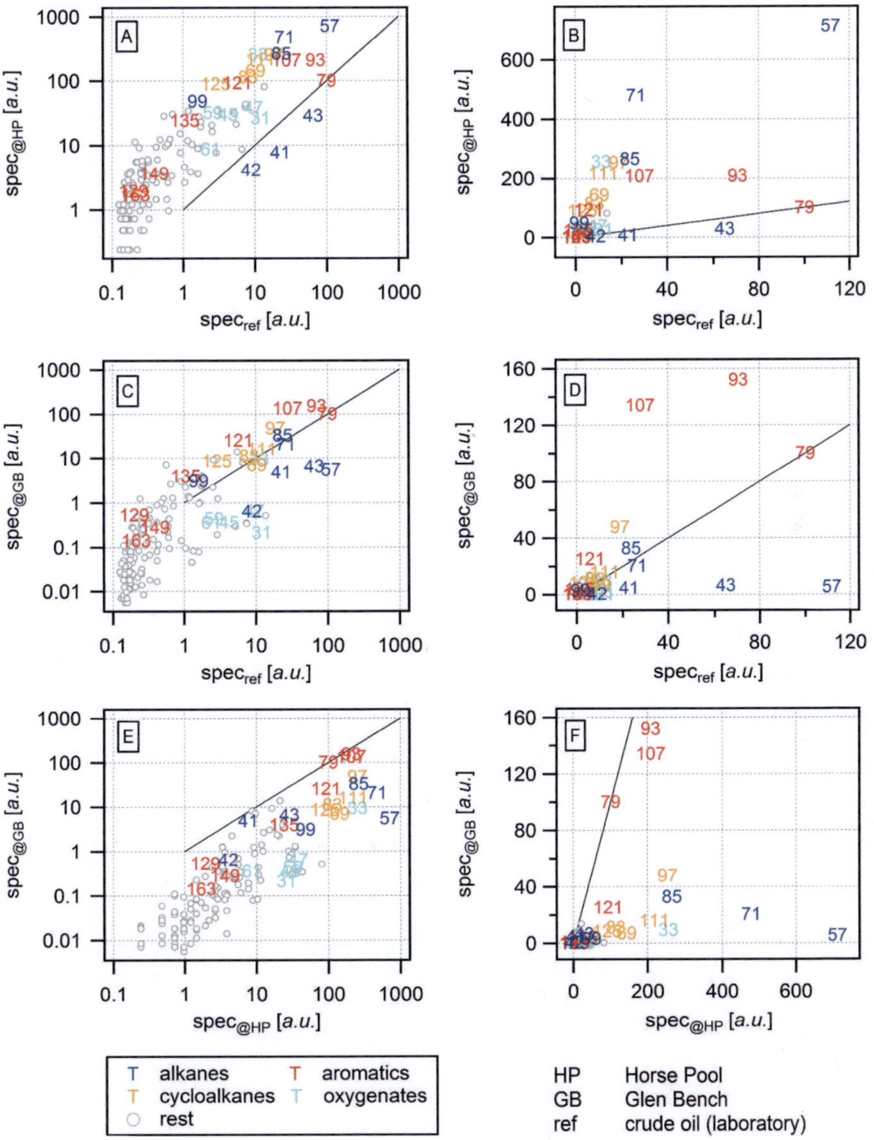

Figure F7.5: *The correlation plots of full mass scans (m/z 20 - 169) are shown in arbitrary units on logarithmic and linear scales;* A/B *represent* spec$_{@HP}$ *vs.* spec$_{ref}$, C/D *represent* spec$_{@GB}$ *vs.* spec$_{ref}$ *and* E/F *represent* spec$_{@GB}$ *vs.* spec$_{@HP}$. *Similar m/z are indicated by their mass numbers and VOC classes are color coded. The 1-to-1 line is only inserted in order to simplify the comparisons between linear and logarithmic projections and has no physical meaning.*

7.4 Comparison of specific well pad emissions

In order to explain the VOC compositions and high background mixing ratios, that were observed at the Horse Pool ground site and during mobile background measurements, emissions at individual point sources were investigated in more detail at oil and gas well pads and other facilities associated with fossil fuel production. The UBWOS 2012 measurement drives using the mobile laboratory created a huge data record, that comprised 38 different gas wells, 12 oil wells, a newly producing well, a recently re-fractured gas well with flow-back pond (→ Glen Bench), and 17 other point sources such as evaporation ponds, storage tanks and compressor stations.

Transect measurements The preferred measurement pattern at point sources were repeated drive bys in close proximity downwind of the emitters/source areas at slow speeds. Special care was taken to the prevailing wind conditions, which ideally were as stable as possible in their respective directions and strengths. GPS tracking together with measured wind fields allowed the precise assignment of emissions to their individual source points on site. Figure F 7.6 (p. 127) shows the drive track at a typical Duchesne County oil well with oil tank, compressor, pump jack and methanol tank color coded according to the concentrations of the emission proxies methane, toluene, methanol and NO_2. A similar illustration is given in figure F 7.7 (p. 128) showing a typical Uintah County gas well with condensate tanks, well head, separator and compressor. In both cases the wind, indicated by wind barbs, was coming from the left hand side, that allowed the detection of the characteristic emissions of the marked point sources on the right hand side downwind from that. Corresponding time series are each shown in the upper panels.

Facility-dependent emissions at production sites Different facilities/systems used for the individual operations on site showed unique emission characteristics. During the transects, such as the one presented in F 7.7 (p. 128), compositions and concentrations were assigned to particular point sources at the well pads and other installations. At most well pads, the condensate tanks (or oil tanks at oil well pads), where produced condensate (or oil) is intermediately stored for later removal, showed distinct emissions of aromatic hydrocarbons (C_6 - C_{13}). Methanol (CH_4O), which was used as anti-freeze and anti-coagulant, showed conspicuous values downwind of the methanol and condensate storage tanks. Methane (CH_4) as one of the principal components of natural gas (simplest existing alkane) showed high values in consequence of direct emissions downwind of the wellheads/ boreholes. Fugitive emissions were also detected at leaking pipelines. Likewise, downwind of separators, where the produced material is split up into liquid and gaseous fractions, high methane values were observed as well. Downwind of compressors, that run on diesel or natural gas in the Uintah Basin, high concentrations of the combustion product NO_2 were found.

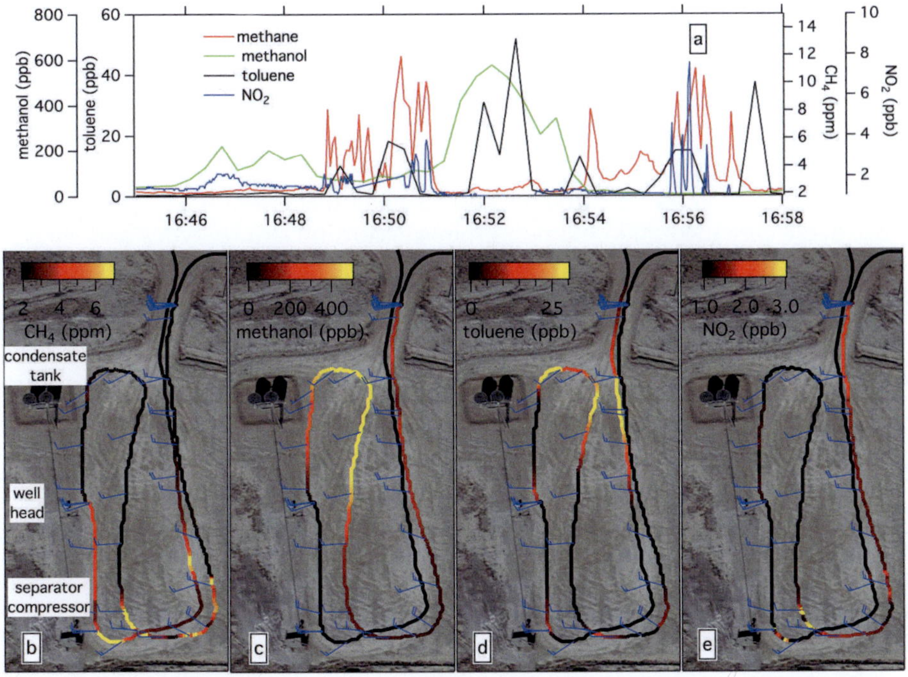

Figure F 7.6: The drive track close to a typical Uintah Basin oil well color coded by measured (b) methane, (c) methanol, (d) toluene and (e) NO_2. Wind barbs indicate the prevailing wind conditions during the measurement. The associated time series are presented in panel (a). (adopted from Warneke et al. *[2014])*

Differences of the examined fields The studied fields in the Uintah Basin, Utah and in Rangely, Colorado, which is another part of the Uinta-Piceance Basin located in the North-West corner of Colorado, differed clearly from each other. Although being only $\sim 30\,km$ apart from each other, the practiced fossil fuel exploration had a very different character. The density of wells close to each other was higher in Rangely, Colorado than in the Uintah Basin, Utah, which allowed adapted practices for the extraction of the natural gas. The important general differences between the investigated regions were (besides the regulatory standards; see section 6.1.1, p. 97): ① gas field type: more wet-gas in the Uintah Basin (→ more liquids) and much drier gas in Rangely, Colorado (→ less hydrocarbon liquids and more gas); ② pipelines: above ground in the Uintah Basin (if used at all) (→ fugitive emissions) and underground to central facility in Rangely, Colorado (→ hardly no on road transportation); ③ powering of well systems: diesel, natural gas in the Uintah Basin (→ NO_x emissions) and electric power in Rangely, Colorado. All these differences resulted in a clear numerical reduction of significant VOC and NO_x emitters (e.g. combustion generators, wellhead compressors, condensate tanks, etc.) and with that of particular emissions in Rangely, Colorado compared to the Uintah Basin, Utah. The mixing ratios for relevant compounds measured in Rangely, Colorado were still elevated from ambient, but generally lower than in the Uintah Basin, Utah.

Classification of emissions due to source characteristics The principal VOC sources were classified into emission groups according similar types and stages of the production procedures, which allowed the comparison of emissions from functional units [Warneke *et al.*, 2014]:

1. **Gas wells** - Uintah County, Utah
    (a) Operator Group A (using **wet-gas** collection - dehydration at central compression facility)
    (b) Operator Group B (using **dry-gas** collection - dehydration on site using glycol dehydrators with extra heater and glycol regenerator)

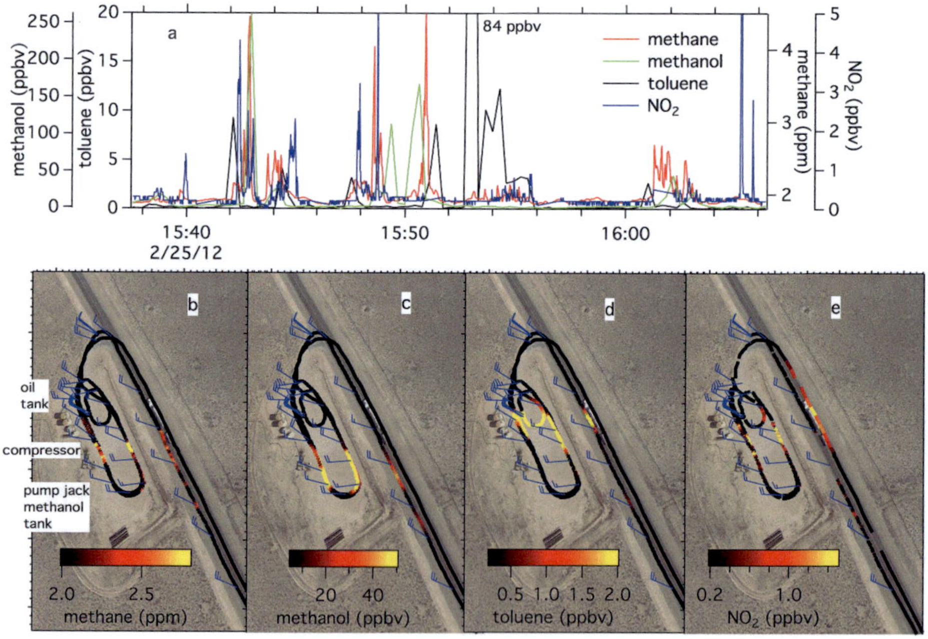

Figure F 7.7: *The drive track close to a typical Uintah Basin natural gas well color coded by measured (b) methane, (c) methanol, (d) toluene and (e) NO_2. Wind barbs indicate the prevailing wind conditions during the measurement. The associated time series are presented in panel (a). (adopted from Warneke* et al. *[2014])*

2. **Oil wells** - Uintah County, Duchesne County (both Utah), Rangely, Colorado
3. **Gas wells** - Rangely, Colorado (as basis for comparisons)
4. A **re-fractured gas well** (Glen Bench) with open flow-back pond (details in section 6.1.5, p. 102)
5. A **newly producing gas well** in the Red Wash fields (close to Glen Bench site)
6. Diverse **other point sources** (including natural gas compressor stations, storage tanks, produced-water injection wells, condensate and oil tank batteries, evaporation ponds and water treatment plants)

The measured mixing ratios were averaged during the time the mobile laboratory spent in close proximity downwind of the wells/emitters, which was generally within a radius $\ll 100\,m$ in order to provide a means of comparison with the Horse Pool site data and aggregated emission inventories. The duration per measurement at one site was usually $20 - 30\,min$ and surveys on the well pads were exclusively performed, when no truck traffic was in the vicinity. Figure F 7.8 (p. 130) shows measurement results for selected compounds. Consider that the averages may be somewhat influenced by the relative amount of background air encountered during the averaging times (i.e. outside of the plumes when the mobile laboratory was turned for the next transect). Nevertheless, the corresponding maximum peak values close to the sources, that are shown in figure F 7.9 (p. 131) for the compounds benzene, methanol and methane, showed similar trends as the averages. This gives indications about the actual enhancements of the mixing ratios close to the sources. Thus, the data can well characterize the sources and allowed their identification. Mixing ratios of aromatics and methanol around $1\,ppmV$ and of methane of hundreds of $ppmV$ were observed several times.

The different source groups showed unique characteristic fingerprints, but even between similar point sources large differences existed. The largest relative disagreements were found between individual gas wells. The highest emissions of aromatic hydrocarbons were observed at the recently hydraulically re-fractured natural gas well Glen Bench (see section 7.5.1, p. 132), followed by the new gas well in the Red Wash fields and the group of dry gas collection wells (operator group B). Likewise, the emissions of cycloalkanes were also most conspicuous for the same source groups. Methanol was observed at most places, but with individually changing concentrations. The average methane mixing ratios at the observation sites showed no clear trends with respect to single groups, but strongly enhanced peak values at Uintah Basin gas wells from operator group B, the new well and the hydraulically fractured well Glen Bench.

Comparison with the WRAP emission inventory The emission inventory provided by the Western Regional Air Partnership[50] (WRAP) gives an analysis of the criteria pollutant emissions for oil and gas exploration and production in the Uintah Basin. The inventory bases on particular production data, well statistics and identified emission sources, that are related to the traced oil and gas source categories. According WRAP about 89 % of the emitted VOCs typically originate from dehydrators, tanks (condensate, oil), pumps and other pneumatic devices and 78 % of the emitted NO_x from drillings rigs, engines (compressors, artificial lifts) and heatings [Warneke *et al.*, 2014].

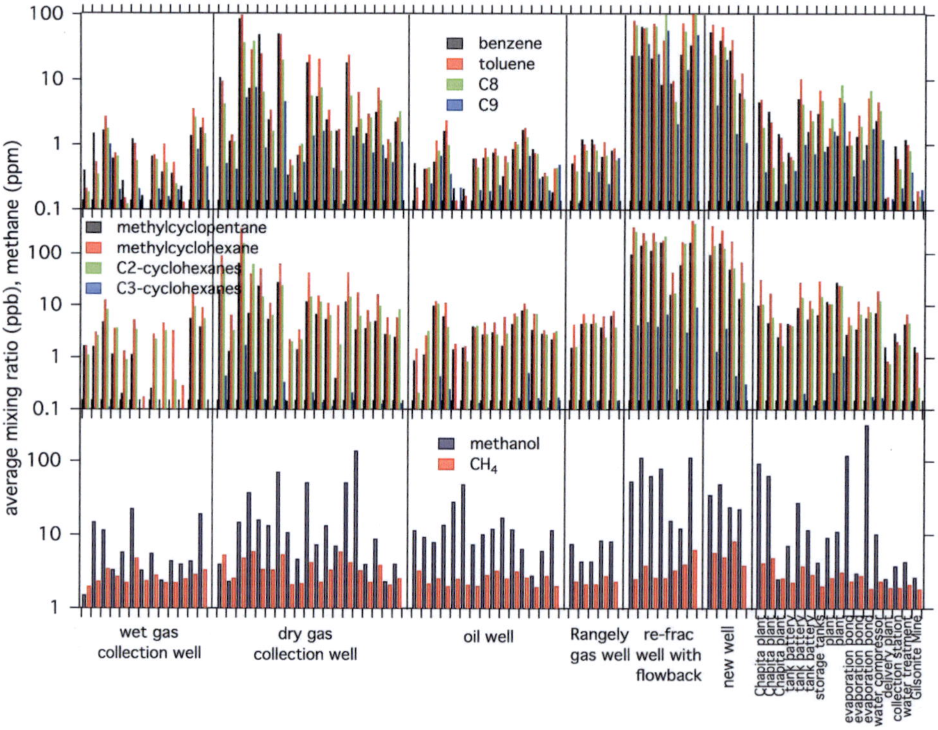

Figure F 7.8: *The average mixing ratios of various compounds (aromatic hydrocarbons C_6 - C_9, cycloalkanes, methanol and methane) measured close to investigated point sources in the Uintah Basin and in Rangely, Colorado. (adopted from Warneke* et al. *[2014])*

[50] available at: `http://www.wrapair.org/forums/ogwg/PhaseIII_Inventory.html`

When comparing the measured VOC compositions with the WRAP emission inventory, it seems that individual wells were the primary contributors to the air mixtures observed in the Uintah Basin. This is in accordance with the observations, because the number of producing oil and gas wells were by far the most numerous point sources in the basin. The VOC compositions at gas well pads were close to conventional gas, but heavier in composition. At oil wells the emissions comprised even heavier compounds than at gas well pads, with the largest difference for aromatic compounds (see also section 7.3, p. 122). The composition of air in the basin would therefore be mixtures of conventional gas (emitted from pneumatic devices, pumps and venting systems) together with oil and condensate tank flashing emissions at oil and gas production sites and emissions from dehydrators (i.e. gas wells with dry-gas collection). The findings of the drives are thus qualitatively similar to what the WRAP inventory affirms.

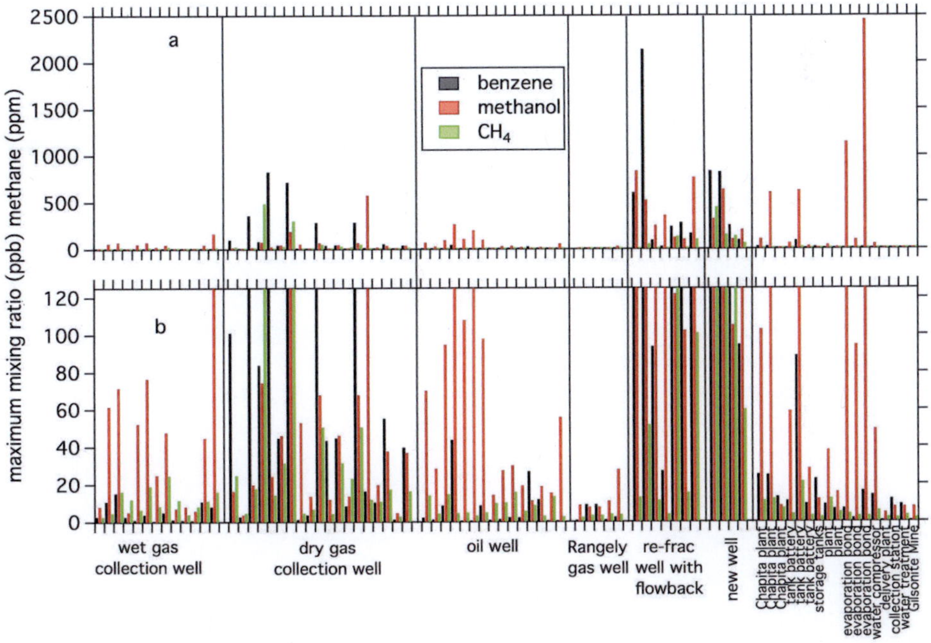

Figure F 7.9: *The maximum mixing ratios of benzene, methanol and methane close to investigated point sources in the Uintah Basin and in Rangely, Colorado in* a *full scale and* b *small scale. (adopted from Warneke* et al. *[2014])*

7.5 Process monitoring: VOC emissions characterization at a flow-back site

The flow-back site Glen Bench was identified to be one of the largest single emitters of organic trace compounds during the performed drives. The well was hydraulically fractured on 9 February 2012 and was officially completed on 11 February 2012. After that day the exploitation and fracturing equipment was reduced on site, while the flow-back phase of the used water into the open pit persisted for subsequent days. All presented measurements took place during this flow-back phase. The reflux into the existing pond continued over the intensive measurement phase of the mission, which officially ended on 1 March 2012.

Although this (re-)fracturing phase with reflux of produced waters into the collection containers generally depicts only a comparatively short time window in the lifetime of an O&NG well, at the same time it represents one of the most substantial exposures of freshly produced materials to the overground environments when using open collection reservoirs.

The site was revisited altogether 14 times on 12 different days at different times of day during the measurement drives in order to monitor variations of the particular emissions connected to the ongoing processes on site. The schedule of the reported visits, which comprised a net measurement time of $\sim 6\,h$, is given in table T 7.1 (p. 133). The processed data consisted of only clearly assignable exhaust air originating from the well site, which is demonstrated in more detail in section 7.5.2 (p. 140). The presented study was performed with the mobile laboratory measuring primarily on the dirt road on the South side of the well, which allowed immediate measurements downwind of the well and the flow-back pond. An illustration of the scenery is given in figure F 7.15 (p. 142) presenting an individual measurement sequence during visit M#10.

7.5.1 Emission peaks of aromatic hydrocarbons

7.5.1.1 Highest observed mixing ratios

In contrast to most other days during UBWOS 2012, a sudden change in the weather towards the most winterly conditions of the complete mission time with snowfall, frozen ground and dense fog characterized the day 19 February 2012. The weather conditions are depicted in figure F 6.6 (p. 111). The measurement drive M#5, which was performed on that day, resulted in the highest observed VOC concentrations during this entire study. In order to illustrate this exceptional period of measured organic trace compounds, figure F 7.10 (p. 134) presents the time series of the corresponding selected aromatic hydrocarbons C_6 - C_{13}.

Measurement Number	Date	Begin [MST]	End [MST]	Duration [hh:ss:mm]
M#1	14 February 2012	19:51:49	20:32:39	00:40:50
M#2	15 February 2012	18:18:31	18:40:11	00:21:40
M#3	16 February 2012	09:27:23	09:56:58	00:29:35
M#4	18 February 2012	19:17:22	19:20:07	00:02:45
M#5	19 February 2012	16:47:30	17:40:50	00:53:20
M#6	20 February 2012	11:46:19	12:13:49	00:27:30
M#7	20 February 2012	18:27:09	19:08:49	00:41:40
M#8	21 February 2012	17:23:44	17:37:04	00:13:20
M#9	22 February 2012	18:38:54	18:56:24	00:17:30
M#10	23 February 2012	17:57:50	18:21:10	00:23:20
M#11	26 February 2012	15:28:52	15:42:12	00:13:20
M#12	27 February 2012	10:21:00	11:01:00	00:40:00
M#13	27 February 2012	17:26:50	17:28:30	00:01:40
M#14	28 February 2012	16:07:22	16:25:42	00:18:20
			$\Sigma =$	05:44:50

Table T 7.1: *The reported VOC investigations performed downwind of the flow-back site Glen Bench with number (M#), date, time [MST] and duration per visit.*

The series display the measured aromatic compounds inside of the plume downwind of the well pad. The first wide peak (16:47 - 17:12) represents initial movings of the mobile laboratory until 17:02 and a subsequent stationary measurement with the van's sensors placed directly in the center of the plume. The following four peaks were connected to two bidirectional transects through the plume both times from West to East and back. Every time the mobile laboratory crossed downwind of the well pad with the flow-back pond, the concentrations of lighter aromatic hydrocarbons peaked to several $ppmV$. The highest mixing ratios were observed for the C_8-aromatics with $\sim 5.2\,ppmV$, but closely followed by toluene, benzene and the C_9-aromatics (all in the $ppmV$-range). Strongest aromatic odors were noticeable on site and characteristic movements on the water surface indicated simultaneous flows of liquids into the pond. Table T 7.2 (p. 135) depicts the observed maximum values for the reported compounds.

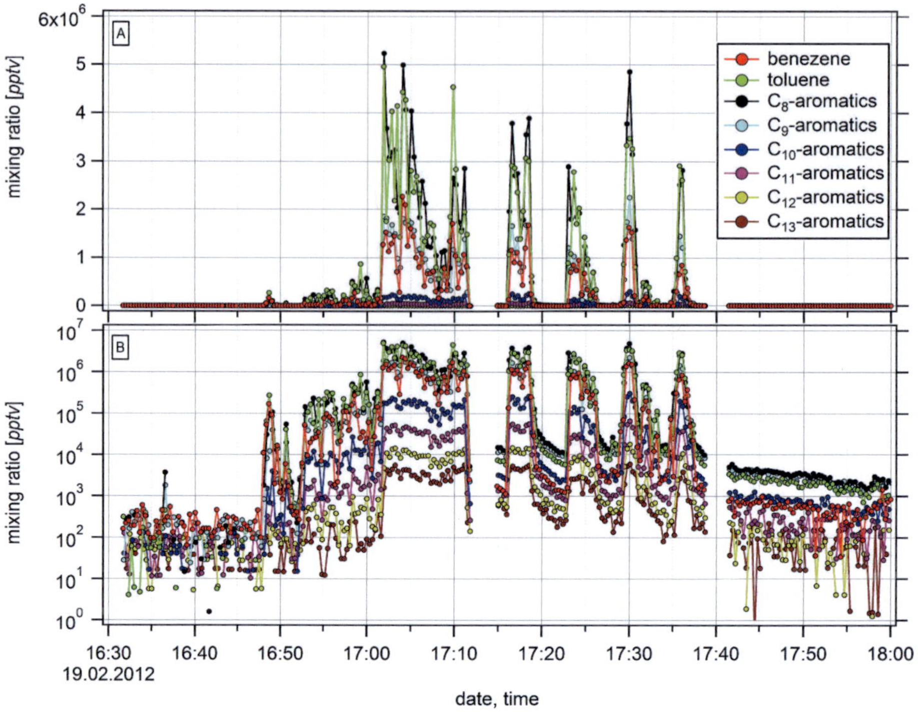

Figure F 7.10: *The time series of C_6 - C_{13}-aromatics (color coded by number of C-atoms) measured downwind of the hydraulically re-fractured well Glen Bench during the mobile laboratory measurement drive M#5 on 19 February 2012. Panel* A *shows linear and panel* B *logarithmic data. Sections without data points are due to system calibrations.*

7.5.1.2 Emission peaks over time

Maximum mixing ratios during the two-week observations The maximum mixing ratios per visit, i.e. independent of individual sources at the well pad, were analyzed for all measurements M#1 - M#14. The measured peak values were ascertained in the described method in average distances of $\sim 60\,m$ to the wellhead, $\sim 30\,m$ to the condensate tanks and $\sim 20\,m$ to the edge of the flow-back pond. A graphic chart showing the temporal changes of the peak mixing ratios of aromatic hydrocarbons C_6 - C_{13}, methanol, NO_2 and the sum of aromatics (i.e. in form of stacked single values) is given in figure F 7.11 (p. 136). High mixing ratios of aromatic compounds were observed over the entire period. Mixing ratios for the lighter compounds benzene, toluene, C_8- and C_9-aromatics reached levels in the $ppmV$-range on multiple days.

Compound	Maximum value [$ppbV$]
Benzene (C_6)	2.3×10^3
Toluene (C_7)	5.0×10^3
C_8-aromatics	5.2×10^3
C_9-aromatics	2.2×10^3
C_{10}-aromatics	306.5
C_{11}-aromatics	66.4
C_{12}-aromatics	18.1
C_{13}-aromatics	5.9
Methanol	1.5×10^3
NO_2	88.3

Table T 7.2: *The maximum mixing ratios of aromatic hydrocarbons, CH_4O and NO_2 at Glen Bench during M#5.*

The weather conditions during all observations changed verifiable from day to day with corresponding horizontal wind speeds on the surface in the range of $1.2 - 10.9\,m \cdot s^{-1}$. In order to evaluate the results of the measurements on a comparable basis accounting for different dilutions at different wind speeds and with that allowing an estimation of the respective source strengths, the data were normalized to a wind speed of $2\,m \cdot s^{-1}$ (by dividing the mixing ratio by the wind speed in $m \cdot s^{-1}$ and multiplying with $2\,m \cdot s^{-1}$) using the average horizontal wind speeds observed during the measurements. The value of $2\,m \cdot s^{-1}$ is arbitrary, but was chosen as standardized basis for these comparisons and finally allowed a conclusion whether the observed differences were either due to different wind speeds or the changing of the source strengths. The resulting series with normalized values are presented in figure F 7.12 (p. 137).

Both representations, observed series and wind normalized series, were characterized by highly variable mixing ratios, which were evidently dominated by the measurements M#5 (19 February 2012) and M#12 (morning of 27 February 2012). The record for the stacked aromatic values was $15.1\,ppmV$, which was obtained during M#5 representing the overall highest observed concentrations. This value decreased to $4.1\,ppmV$ after the wind normalization, as the horizontal wind speeds at that time were rather high. In return to that, measurement M#12 with observed stacked aromatic mixing ratios of $5.4\,ppmV$, showed a distinct enhancement after wind normalization to the new peak value of $7.9\,ppmV$.

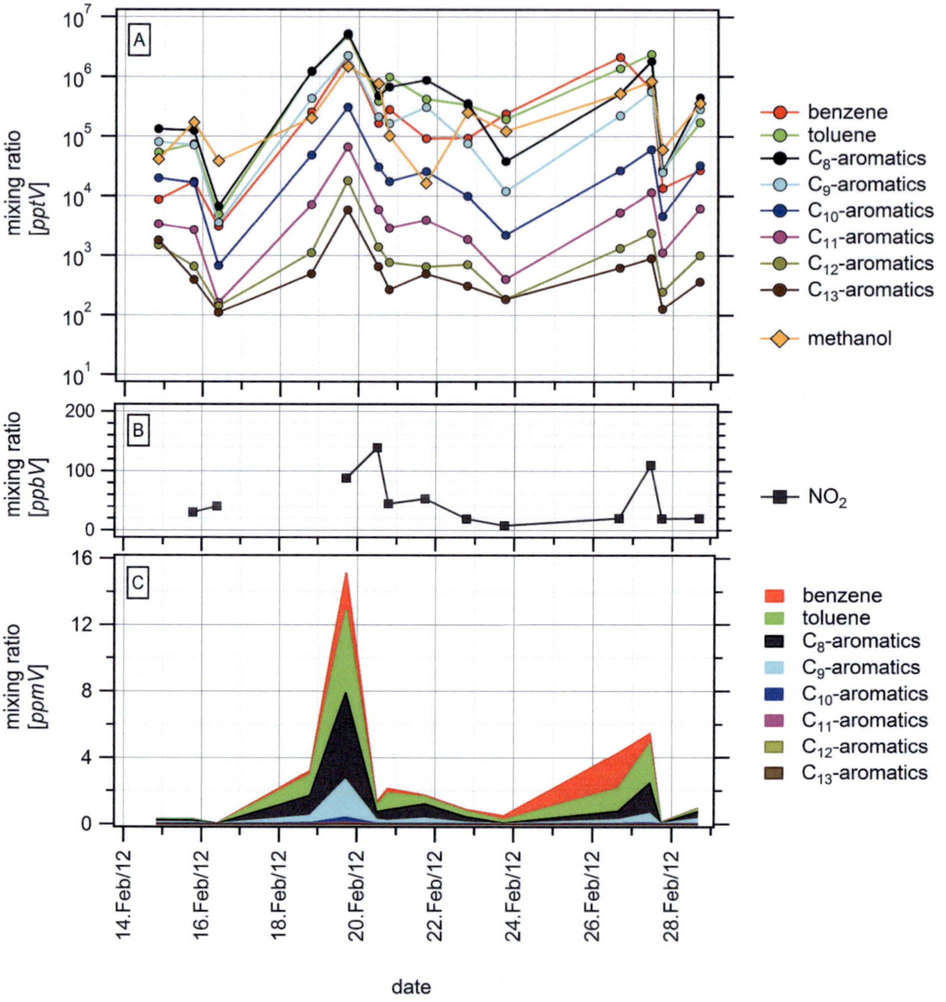

Figure F 7.11: *The observed maximum mixing ratios for M#1 - M#14 at the Glen Bench flowback site with panel* A *showing the aromatic hydrocarbons (C_6 - C_{13}) and methanol in [pptV] and panel* B *showing the maximum NO_2 mixing ratios in [ppbV]. The stacked single values C_6 - C_{13} are presented in panel* C *in [ppmV].*

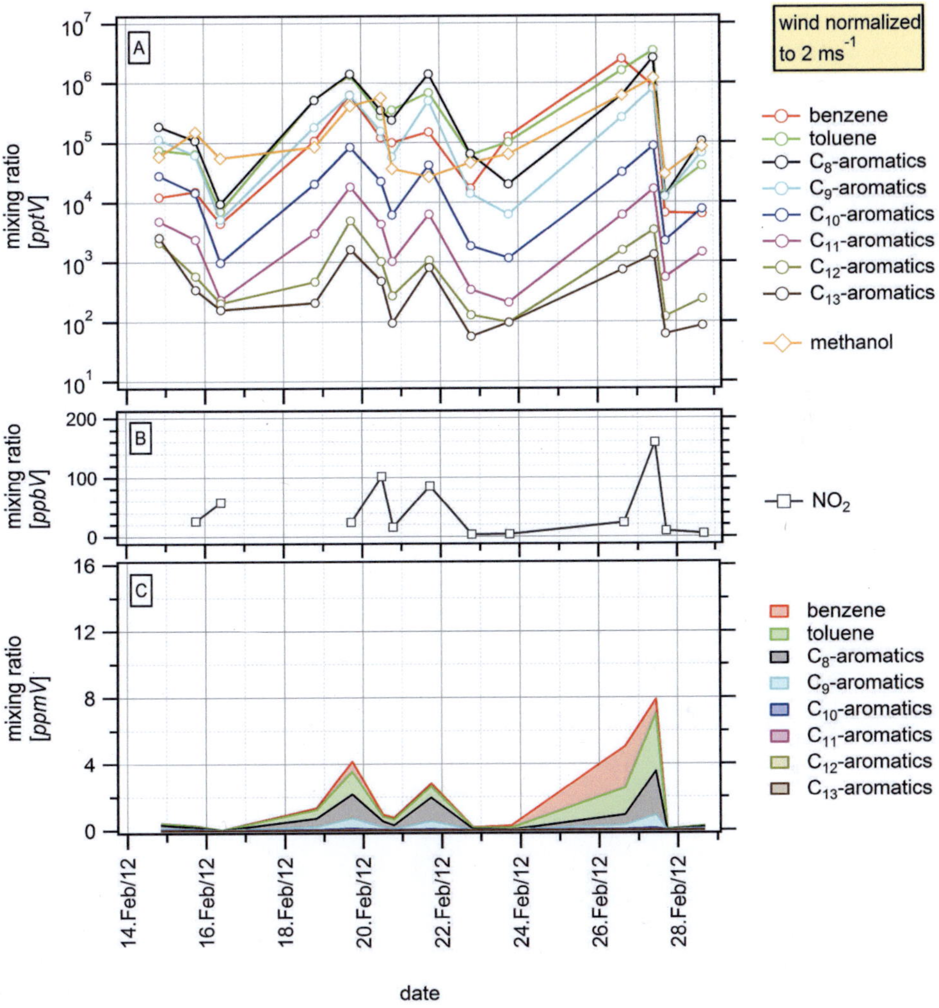

Figure F 7.12: *The wind normalized (to* $2\,m\cdot s^{-1}$*) maximum mixing ratios for M#1 - M#14 at the Glen Bench flow-back site with panel* A *showing the aromatic hydrocarbons (*C_6 *-* C_{13}*) and methanol in [pptV] and panel* B *showing the maximum* NO_2 *mixing ratios in [ppbV]. The stacked single values* C_6 *-* C_{13} *are presented in panel* C *in [ppmV].*

No clear trend towards a general in- or decrease of maximum emissions of aromatic compounds was found for the investigation period for both directly measured and wind normalized concentrations, which indicated that processes on site determined the observed mixing ratios. As the flow-back phase with the open pit was not ended during the presented measurements, a final conclusion concerning the emission development towards the normal production phase was not available.

7.5.1.3 Percentage compositions of the aromatic hydrocarbons

The aromatic compounds did not show clear trends in measured concentrations within the observation period. Thus, individual fluctuations in their percentage compositions over time were determined in order to assess internal processing. Corresponding distributions of fractional amounts of C_6 - C_{13} to the total aromatics are presented in figure F 7.13 (p. 138).

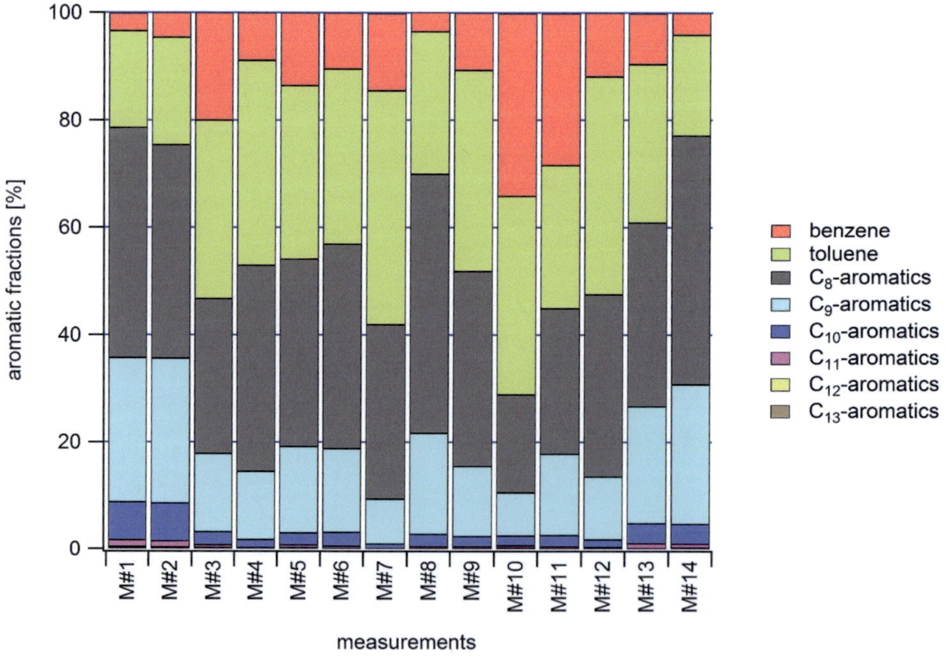

Figure F 7.13: *The relative fractions of aromatic hydrocarbons (C_6 - C_{13}) to the total signals observed downwind of the well Glen Bench during the mobile laboratory measurement drives.*

These proportions were derived using integrated VOC enhancements during downwind periods in order to cover the complete spectrum of emissions coming from the well pad (including condensate tanks and flow-back pond, etc.). The reason for this approach attributed to the fact, that otherwise measurements with superposing signals as consequence of crosscutting wind directions would have been excluded from this analysis. This was important, since most measurements were affected by wind situations, that allowed an overlapping of trajectories from different point sources on the well pad. The only exception from that, which could be observed during the measurements, is presented in figure F 7.15, p. 142.

The C_8-aromatics were the largest fraction in more than half of all investigated measurings covering the entire processes at the flow-back site. They represented even $\sim 48\,\%$ of the total composition of aromatic compounds during M#8 (21 February 2012). Moreover, the C_8-aromatics exceeded the percentage of the sum of the other higher aromatic hydrocarbons C_9 - C_{13} in all investigated periods. The smallest fractions of C_8-aromatics were observed on days with the largest fractions of benzene, but no simple counterbalancing was observed between these two. Benzene and toluene correlated well during all measurements. All other compounds C_8 - C_{13} on the other hand generally anti-correlated with this pair.

In order to further evaluate the development of internal processes within the compositions, enhancement ratios (ER) of all compounds with benzene were used. The corresponding ER time series are presented in figure F 7.14 (p. 140). In addition the average benzene mixing ratios per measurement are enclosed to the respective ER values in order to illustrate the differences to the peak values (as shown above). Likewise the 1-to-1 line, which basically represents the ER of benzene with benzene and therefore depicts the positive or negative enhancement threshold, is included. The highest ratios were found for the C_8-aromatics with peak values in the beginning (M#1 - 15 February 2012), the middle (M#8 - 21 February 2012) and the end (M#14 - 28 February 2012) of the observation period. The maximum value was 13.6 during M#8.

Neither the proportional compositions nor the ER showed clear directional trends in any way, which would have been expected for undisturbed systems. This lack of indications for a clear direction within the internal processing of the aromatic compositions together with the determined fluctuations in the total aromatic mixing ratios provided evidence, that external influences/processes changed the measured VOC compositions between the single measurements. Thus, individual well operations and specific peculiarities of the flow-back site (e.g. interplay of several different point sources at the well pad) have to be assumed to be responsible for these effects.

A further analysis of the aromatic composition measured at Glen Bench is presented in section 7.5.2 (p. 140) for a measurement allowing the separate investigation of decoupled plumes from tanks and pond. Associated flux estimations are shown in section 7.5.3 (p. 146).

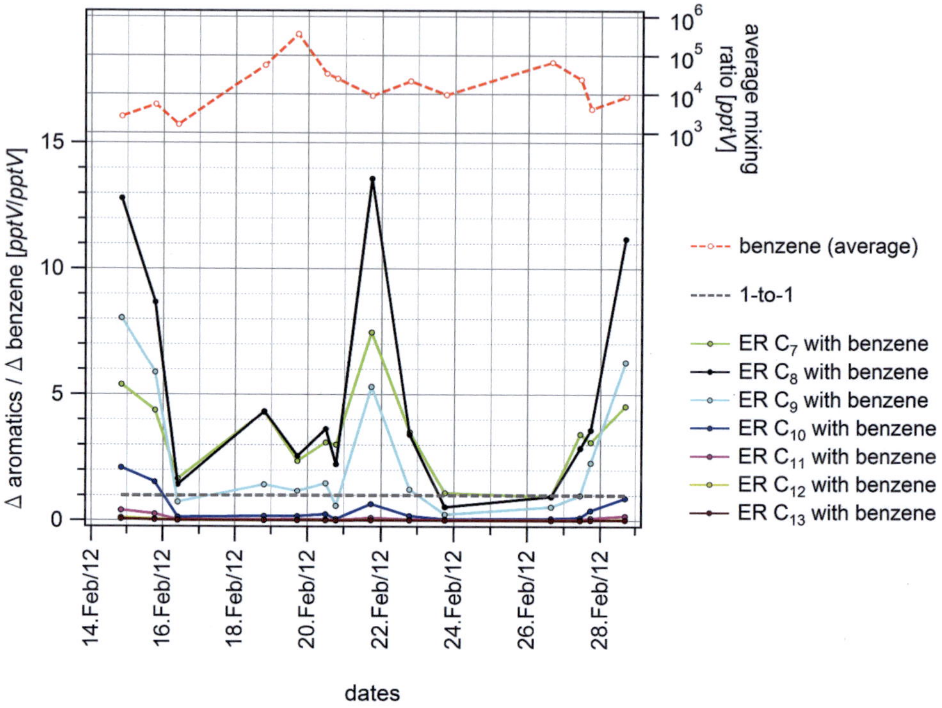

Figure F 7.14: *The enhancement ratios (ER) for all reported aromatic compounds with benzene are shown in* $^{\Delta aromatics}/_{\Delta benzene}$ *(*$[^{pptV}/_{pptV}]$*) together with the averaged benzene mixing ratios in [pptV].*

7.5.2 Differentiation of emission sources

A measurement sequence of an immediate crossing downwind of the well pad at the Glen Bench site during M#10 enabled a detailed characterization of the individual point sources on site. This allowed the differentiation between the plumes from condensate tanks and flow-back pond. Figure F 7.15 (p. 142) presents this bidirectional crossing of the plumes in form of georeferenced trace gas profiles of the compounds toluene, benzene and C_8 - C_{10}-aromatics, methane and NO_2 with color coded mixing ratios superimposed to a topographic map of the area. Wind barbs show the wind conditions at that moment and subsystems on the well pad are highlighted.

The drive track consisted of a loop with 3 side arms, that chronologically followed in the order indicated by the reference points. These points additionally display locations used for dispersion model calculations of fluxes, which are discussed in more detail in section 7.5.3, p. 146. The clear differentiation of the different plumes was obtainable due to the favorable wind conditions during that measurement. On other days comparable results were found as well, but not as clearly allocatable. Reasons for this were more turbulent wind fields and crosscutting wind directions during these other measurements, that complicated the precise differentiation of the plumes and with that a clear assignment of detected signals to sources.

Identification of source areas In order to highlight temporal coincidences, the time series of the presented compounds are juxtaposed in figure F 7.16 (p. 143). Colored boxes represent periods that enabled the exclusive detection of air originating from condensate tanks or flow-back pond. The moments, when the sample air additionally contained the plumes from the other subsystems on the well pad, are indicated by the occurrences of the proxies CH_4 (originating from condensate tanks, wellhead and separator), methanol (originating from the methanol canister) and NO_2 (originating from the generator-powered separator). The corresponding average mixing ratios for the aromatic compounds are given in table T 7.3 (p. 144) showing the existence of two aromatic domains with distinctly different compositions. The lighter compounds C_6 - C_8 dominated the air mixture that was emitted from the condensate tanks, and the heavier compounds dominated the mixture from the open flow-back pond. An illustration of this contrasting is presented in figure F 7.17 (p. 144) showing the different distributions with particular shifting of the curve maxima between both emitters. This showed, that tanks and pond were characterized by clearly different VOC fingerprints, that contributed in different ways to the measured total signals downwind of the well pad of the Glen Bench site.

The percentages of individual aromatic compounds to the sum of all aromatic species for the two emission domains tanks and pond are shown in figure F 7.18 (p. 145). The pie charts represent averages from the transects presented in figure F 7.16 (p. 143) and particularly show that the condensate tank emissions were composed of 46 % from benzene, 42 % from toluene, 11 % from C_8-aromatics and to a negligible rest of together $\sim$ 1 % from C_9 - C_{13}-aromatics. In contrast to that, the flow-back pond emissions were more diversified and were composed of 10 % from benzene, 27 % from toluene, 32 % from C_8-aromatics, 23 % from C_9-aromatics, 6 % from C_{10}-aromatics and to a rest of together $\sim$ 2 % from C_{11} - C_{13}-aromatics. By comparing with the integrated signal (with 34 % from benzene, 37 % from toluene, 18 % from C_8-, 8 % from C_9-, 2 % from C_{10}-aromatics), that corresponds with the fractions of aromatic compounds shown in figure F 7.13 (p. 138), it was evident that the larger contributions of lighter compounds (i.e. benzene and toluene) originated from the condensate tanks and the heavier compounds from the flow-back pond (i.e. C_8 - C_{13}-aromatics).

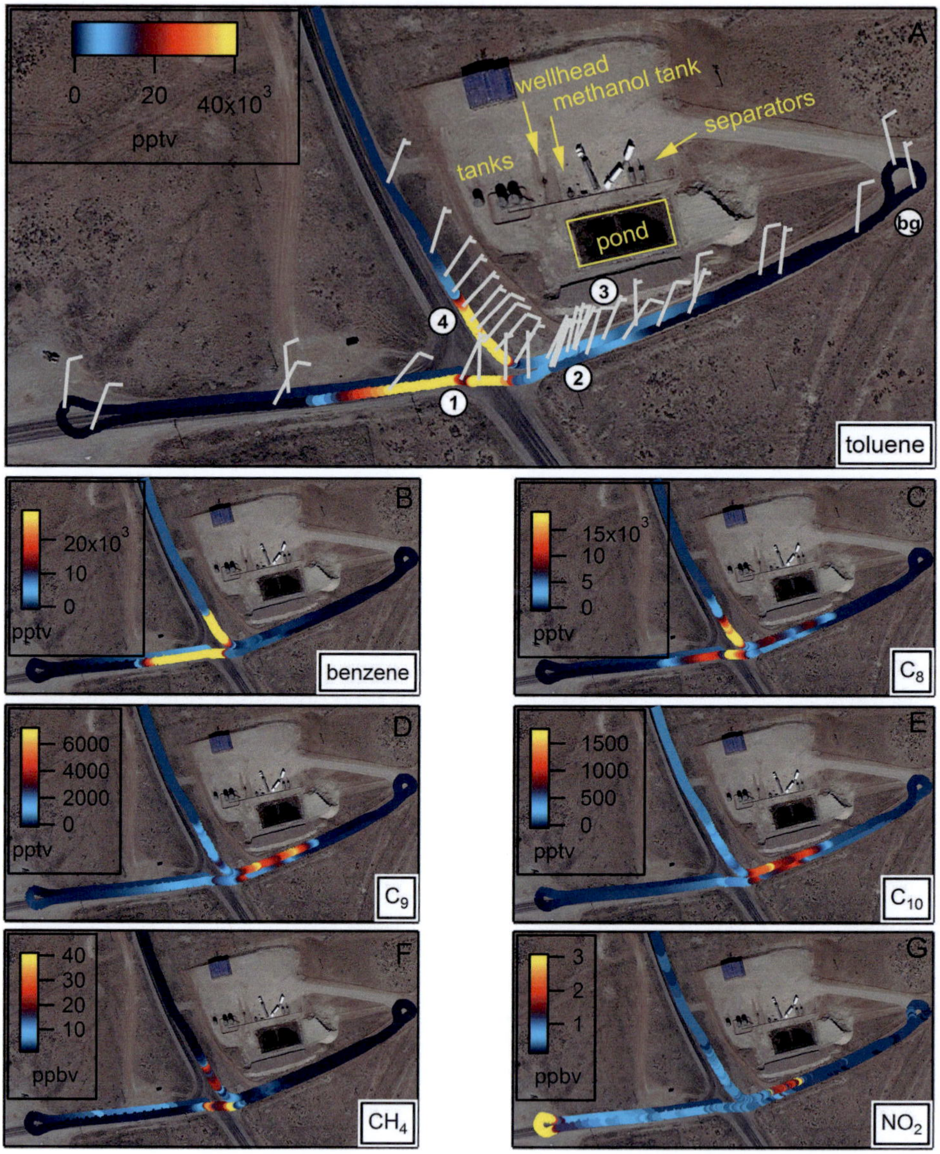

Figure F 7.15: *The satellite picture of Glen Bench (adopted from Google Earth; imagery date: 22 March 2012) is overlayed with a color coded drive track of the mobile laboratory illustrating different compounds measured during M#10 (23 February 2012). In the upper panel various subsystems on the well pad, simulation reference points ① - ④ and a background point (bg) are additionally highlighted. Wind barbs indicate the prevailing wind fields at that moment.*

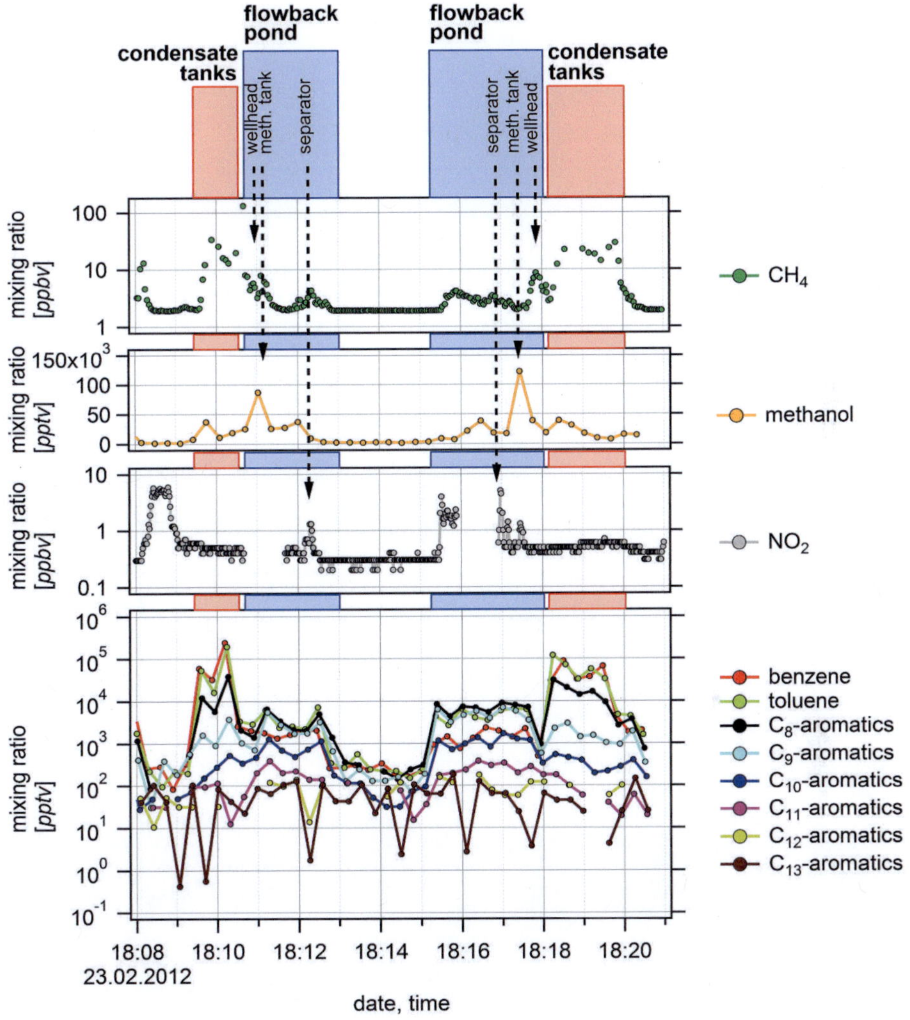

Figure F 7.16: *The color coded time series for methane (in [ppbV]), methanol (in [pptV]), NO_2 (in [ppbV]) and the aromatic hydrocarbons (C_6 - C_{13}) (in [pptV]) are shown for a bidirectional transect downwind of the Glen Bench site during M#10 (23 February 2012).*

Compounds	averaged mixing ratios in [*ppbV*] downwind of **condensate tanks**	**open flow-back pond**
Benzene (C_6)	80.4	1.7
Toluene (C_7)	75	4.6
C_8-aromatics	18.6	5.5
C_9-aromatics	2	4
C_{10}-aromatics	0.34	1
C_{11}-aromatics	0.071	0.24
C_{12}-aromatics	0.03	0.09
C_{13}-aromatics	0.038	0.07

Table T 7.3: *The average mixing ratios of the aromatic compounds C_6 - C_{13} measured downwind of the condensate tanks and open flow-back pond at Glen Bench during M#10 (23 February 2012).*

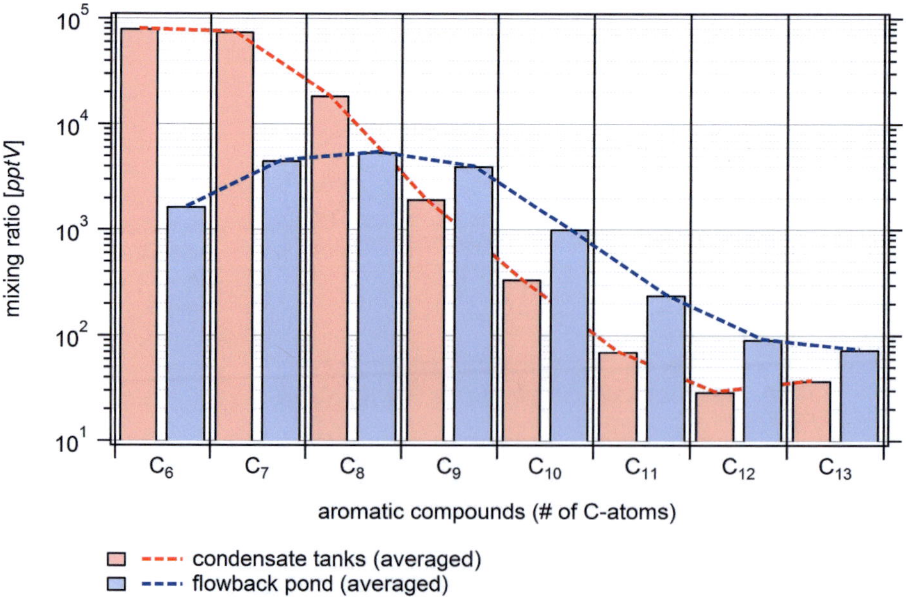

Figure F 7.17: *The grouped bar chart of averaged mixing ratios of aromatic compounds per number of C-atoms in [pptV] for condensate tanks (red) and flow-back pond (blue) as measured during M#10.*

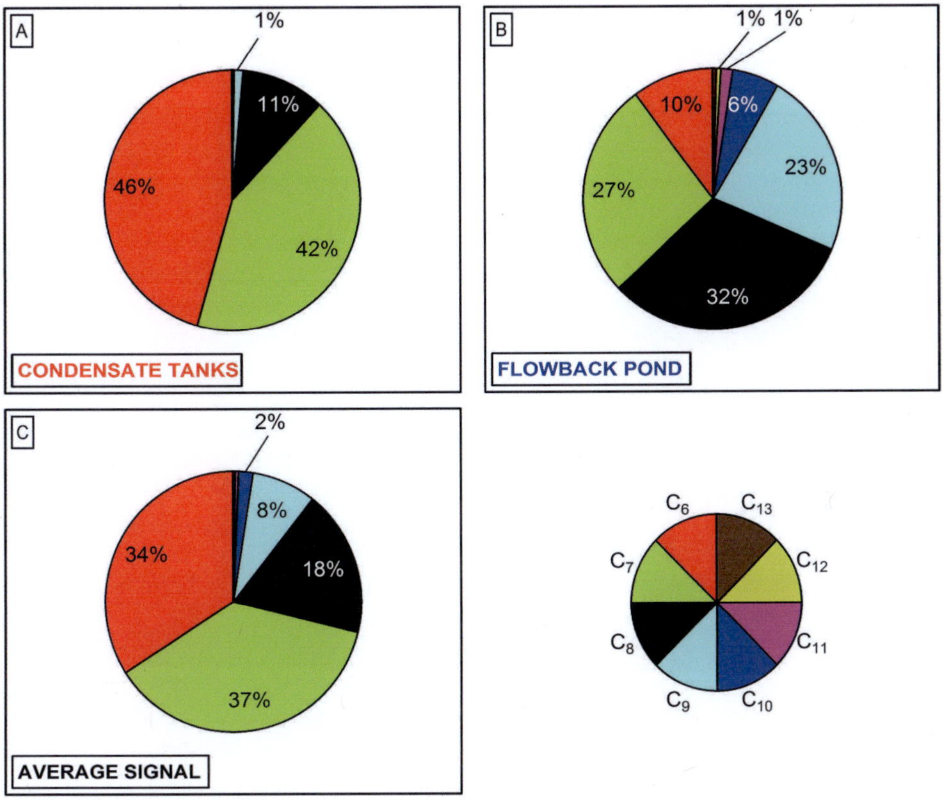

Figure F 7.18: *The percentage composition of aromatic compounds measured downwind of condensate tanks and flow-back pond during M#10: panel* A *condensate tanks, panel* B *flow-back pond and panel* C *integrated signal as used in F 7.13 (p. 138).*

Conclusions from the observations On the basis of this clear differentiation between individual emissions from condensate tanks and flow-back pond, it was concluded that both emitters were to be classified as different VOC sources with independent compositions. Thus, the total signals measured downwind of the Glen Bench site during the measurements presented in section 7.5.1 (p. 132) should be understood as compositions particularly affected by differently weighted contributions from tanks and pond. As the groups of lighter and heavier aromatic compounds originating from the two emission domains were additionally observed to be anti-correlated among each other (see section 7.5.1.2, p. 134), a possible dominance of either of the two groups could be concluded for the presented measurements. This directly pointed to a connection of the measured individual compositions with processes happening at the well site (e.g. filling schedule into tanks or pond; varying produced water compositions; etc.).

Possible explanations for the formation of the two domains were the handling of the produced materials, that were separated into tanks (possibly via the separator) and pond due to the fractions of (diluted) oil and gas in the produced waters, and enhanced evaporation of lighter aromatics from the pond. This latter effect has been reported by Li *et al.* [2013], who demonstrated that lower volatility compounds (mostly associated with heavier compounds) evaporate more slowly and delayed than higher volatility compounds (mostly associated with lighter compounds). Evaporation rates from crude oil depend on the vapor pressures of the compounds together with other conditions like temperatures, reaction times and surface areas and vary for different VOC species [de Gouw *et al.*, 2011; Yuan *et al.*, 2014]. Likewise, enhanced dissolving of lighter aromatics in the water column prior to surfacing has been reported by de Gouw *et al.* [2011]; Middlebrook *et al.* [2012]; Ryerson *et al.* [2011, 2012] for the case of oil being flown out into the ocean after the explosion of the oil platform Deepwater Horizon in the Gulf of Mexico. All these effects possibly influenced the processing of the produced materials in both repositories in different ways and most likely explain the observed effects.

7.5.3 Flux estimations using a Lagrangian dispersion model

The presented observations of VOCs in the Uintah Basin based on actual in situ field measurements. With the derived wind normalized VOC mixing ratios (see figure F 7.12, p. 137) equivalents for source strengths of compounds originating from the Glen Bench site were determined. However, these performed measurements did not account for the vertical properties of the individual plumes (i.e. types and shapes) and therefore did not enable the direct deduction of emission rates of the single emitters.

The simple atmospheric Lagrangian dispersion model WindTrax2.0[51] for statistical simulations of forward and backward trajectories of turbulent short-range transport was used for the estimation of emissions in consideration of typical plume characteristics. For the presented model runs standard atmospheric parameters in connection with geographic reference settings for Glen Bench, i.e. an altitude of $1716\,m$ above sea level and surface roughness constraints compatible with on-site observations, were applied. The simulations comprised 1 million test particles per run. Some more detailed descriptions of the backward Lagrangian Stochastic model itself can be found in Flesch *et al.* [1995] and instructions for comparable applications are given in Crenna *et al.* [2008]; Flesch *et al.* [2005].

The Lagrangian model was used to estimate the fluxes of aromatics at Glen Bench exemplary for M#10, whose particular situation, as shown in figure F 7.15 (p. 142),

[51] `http://thunderbeachscientific.com/`

consisted of uncoupled plumes originating from condensate tanks and flow-back pond. The proxy toluene (C_7) was used for this estimation and corresponding mixing ratios, as well as the average wind field, were applied to the model. Because of simplification and in order to avoid an over-determination of the scenery, the emissions of the tanks (standard condensate tanks and catch tank) were combined. Likewise, a static underlying background concentration was derived using toluene average background values (background mixing ratio: $0.283\,ppbV$). 3D-locations of the sources, i.e. tanks and pond, as well as of sensors, i.e. mobile laboratory, were applied to the model due to georeferenced allocations on the satellite picture, GPS tracking data and on-site observations. The underground was set to be flat. Corresponding reading points for the four presented reference positions and the background measurement are depicted in the aforementioned figure as well, but for the sake of clarity next to their actual positions on the color coded track.

The calculated emission rates for the condensate tanks under the stated simplifications are presented in table T7.4 (p. 148). The toluene emission rates for the condensate tanks were calculated for the two different positions ① and ④ and agreed well with $\sim 0.15\,g\cdot s^{-1}$.

In contrast to these clustered point sources, the flow-back pond had a huge surface source area. By evaluating of the satellite picture, the surface of the waste water pond was determined to be $\sim 730\,m^2$. The calculated toluene emission rates for the positions ② and ③ downwind of the flow-back pond are given in table T7.5 (p. 148). They agreed well with $\sim 1.9 \times 10^{-5}\,g\cdot m^{-2}\cdot s^{-1}$.

In consideration of the extensive surface area, integrated emission rates resulted in a total value of $\sim 0.014\,g\cdot s^{-1}$, which was roughly one order of magnitude smaller than the emissions from the tanks. This relationship also agreed in a zeroth order approximation with the primary in situ observations of the mixing ratios made during the measurement drive (see section 7.5.2, p. 140). Thereby the average toluene mixing ratios were determined to be $\sim 75\,ppbV$ downwind of the tanks and $\sim 4.6\,ppbV$ downwind of the flow-back pond, which also differed roughly by one order of magnitude. Although average mixing ratios are not qualified to replace fluxes as a matter of course, at least their proportionality was reproduced for this example. This indicated, that the applied modeling method in combination with the measured transects at well pads was suitable to evaluate emission scenarios qualitatively and vice versa, that the wind normalized mixing ratios were a good first impression for the magnitudes of source strengths when comparing emission scenarios such as the presented situation. However, as there were no comparable studies in the literature that looked at VOC emission rates at other hydrofracking sites, the derived values were not verifiable using other comparable results. Thus, the presented estimations were classified as case study fitting exclusively to the unique situation of M#10 under the described simplifying assumptions.

Further comprehensive investigations especially of the plume characteristics are needed for a more detailed estimation of fluxes or source strengths, respectively. This is also necessary for more complex wind scenarios with superimposed mixing ratios of tanks and pond due to intersecting plume trajectories. For determining those emission fluxes from well pads, techniques like tracer releases, remote geospatial Gaussian methods or the precise measurement of individual equipment parts on the well pad would be suitable.

Tanks	**Emission rate** $[g/s]$	**Standard deviation**
Position 1	0.148	3.42e-3
Position 4	0.154	3.189e-3

Table T 7.4: *The toluene emissions from the condensate tanks estimated using the atmospheric Lagrangian dispersion model WindTrax2.0 for a sequence measured during M#10.*

Flow-back Pond	**Emission Rate** $[g/m^2/s]$	**Standard Deviation**	**Area-integrated Emission Rate** $[g/s]$	**Area-integrated Std. Dev.**
Position 2	1.89e-5	1.158e-6	1.381e-2	8.463e-4
Position 3	1.911e-5	1.907e-6	1.396e-2	1.393e-3

Table T 7.5: *The toluene emissions from the flow-back pond estimated using the atmospheric Lagrangian dispersion model WindTrax2.0 for a sequence measured during M#10.*

7.6 Hydrogen sulfide (H_2S) measurements

Hydrogen sulfide (H_2S) is a highly toxic, strongly smelling and highly flammable gas, that can be detected at various locations mostly connected to volcanic activities, natural decompositions of sulfates and sulfur-containing organic compounds by anaerobic bacteria or routine/accident industrial releases. Likewise, its occurrence at fossil pits, landfills, livestock manures and other polluted waters and areas has been verified. [ATSDR, 2006; EPA, 1993; Li *et al.*, 2014; Tarver & Dasgupta, 1997]

One of the primary atmospheric sources of H_2S is associated with O&NG operations, where this normally unwanted by-product is a serious danger for workers and local residents at production sites.

Since hydrogen sulfide is not a classic VOC, PTR measurements of this compound had not been considered in this context before. The idea to investigate H_2S came up during the drives, as relevant concentrations in the basin's air were expected to be present and appropriate warning signs at various well pads were found. In order to identify and map the H_2S distribution in the Uintah Basin, the ULW-PTR-MS measured the associated detection mass at m/z 35 [Feilberg *et al.*, 2010; Liu *et al.*, 2011] for a certain period. Necessary preparatory and post-treatment works for a scientific assessment of the measured hydrogen sulfide signals were performed by Li *et al.* [2014] and are described in more detail in section 4.1.3 (p. 68).

Four measurement drives probing hydrogen sulfide Figure F 7.19 (p. 150) shows the obtained hydrogen sulfide signals for the four measurement drives with attention to this compound. The drives headed for different regions. On 25 February 2012 the mobile laboratory performed measurements in the oil fields in Duchesne County, Utah. The following two drives on 26 and 27 February 2012 were concentrated to the Northern and Southern parts of the Uintah County, Utah gas fields. The last drive with special attention to H_2S on 28 February 2012 was conducted to the oil and natural gas fields in Rangely, Colorado, but covered some areas around the coal-fired power plant at Bonanza, Utah on the way back to Vernal, Utah as well. In the presented figure the drive tracks each show the corresponding signals of H_3S^+. Color and size coded circles indicate the significant presence of hydrogen sulfide. Most locations showed small levels below $50\,cps$ representing background values, with signal enhancements at particular locations with factors of up to ~ 3. Nonetheless, at some places explicitly higher signals were found. One location even showed m/z 35 signals of $\sim 40'000\,cps$, which is an enhancement of 3 orders of magnitude. The relevant site was near a low-laying sink in the Southern gas fields at a condensate tank, that was under service at that moment. The location is highlighted in the figure with pink circles.

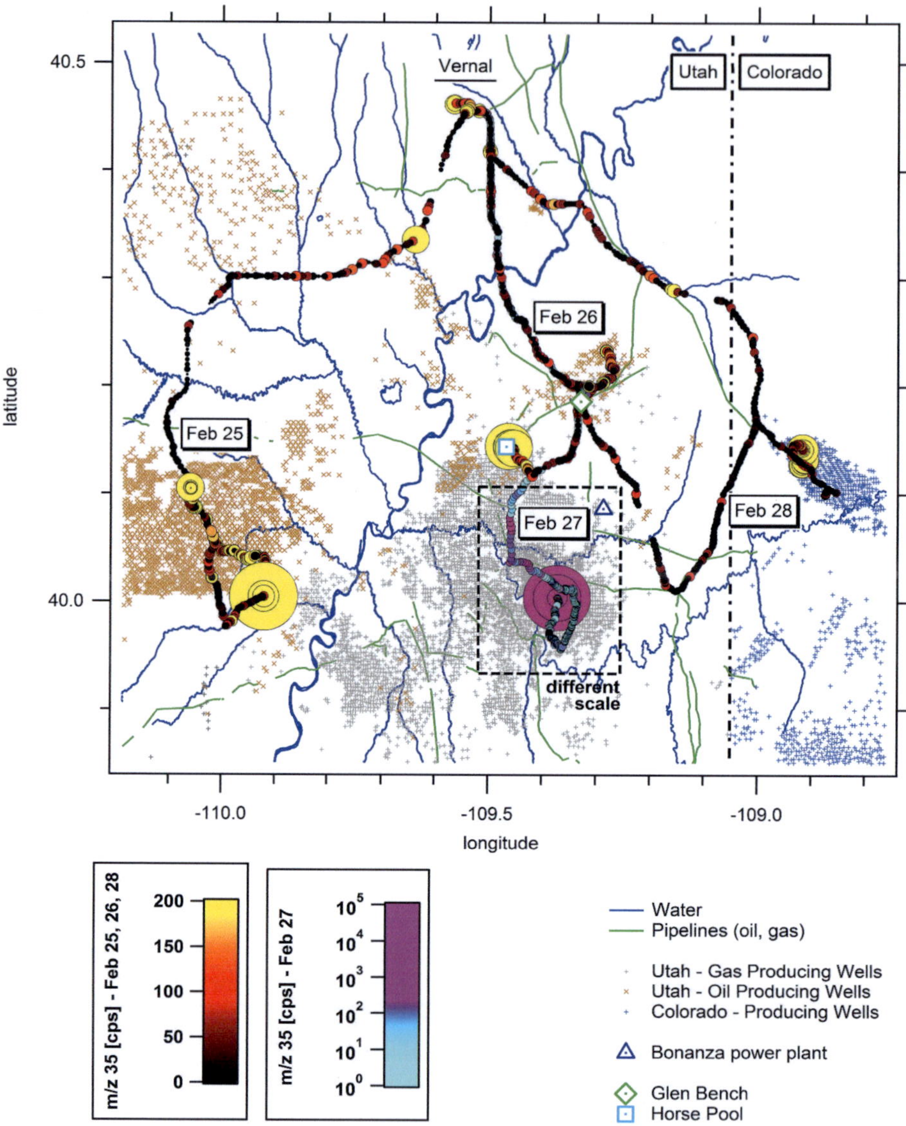

Figure F 7.19: *Mobile laboratory tracks color and size coded with H_2S signals at m/z 35 among the oil and natural gas wells and other O&NG infrastructures in the Uintah Basin during UBWOS 2012. The four drives (25, 26, 27, 28 February 2012) are not directly comparable with respect to circle sizes as they are not synchronized in that representation.*

In consideration of the aforementioned retroactive calibration work, it can be assumed that H_2S mixing ratios were at $4 \pm 2\,ppbV$ at most areas in the O&NG fields close to well pads due to routine emissions from oil and gas facilities. This represents clearly elevated values compared to ambient air mixing ratios of $0.02 - 0.07\,ppbV$ in undeveloped areas and $0.11 - 0.33\,ppbV$ in urban areas [ATSDR, 2006; Skrtic, 2006]. Significant H_2S releases at unique point sources, like e.g. the described servicing at the tank, showed mixing ratios of up to $9 \pm 4ppmV$.

Although representing only rough estimations of H_2S occurrences and mixing ratios in the Uintah Basin, the presented measurements demonstrated for the first-time the applicability of PTR instruments for mobile fast-response detections of hydrogen sulfide. This directly contributes to future PTR-MS measurements of emissions in the context of O&NG production, as it enables the parallel real-time detection of this indicator for hazardous emissions without the need for additional instruments.

Summary and Outlook

This work reports on the development, characterization and use of the Ultra-Light-Weight Proton-Transfer-Reaction Mass Spectrometer (ULW-PTR-MS) and studies about emissions of organic trace compounds from oil and natural gas production sites during air quality investigations in a mountain basin in the Western United States.

The ULW-PTR-MS is a new high-sensitivity in situ instrument for fast-response measurements of VOCs in ambient air. It has been developed to be deployed aboard the German research aircraft HALO. The home-built setup is characterized by its compactness and little weight of $\sim 55\,kg$ compared to commercial systems with typical weights of $> 130\,kg$. The developed control electronics with associated software as core element of the system architecture base on a modular measurement and control system, that allows fast adaptations to mission-specific demands. The final construction fulfills the engineering and safety requirements due to the aeronautical certification and reliably works under the harsh operating conditions during airborne measurements with strongly varying ambient pressures at altitudes from sea level to the ceiling cruise altitude of HALO ($15'500\,m$).

The ULW-PTR-MS was first-time used for field measurements during the Uintah Basin Winter Ozone Study 2012 (UBWOS 2012) in Utah/USA, which was an air quality study investigating the chemistry of ozone and its precursors in a region of intense fossil fuel production. Due to the initial parallel deployment to other high-sensitivity VOC instruments from the National Oceanic and Atmospheric Administration (NOAA) in Boulder, Colorado/USA at a field site in the Uintah Basin, Utah, immediate comparisons and calibration measurements attesting the accuracy of the ULW-PTR-MS were carried out. The following deployment on board of a laboratory van performing several weeks of measurement drives across oil and natural gas fields enabled the identification of emissions from individual oil and natural gas well pads and other point sources associated with fossil fuel production. High mixing ratios of VOCs (alkanes, cycloalkanes, aromatics, oxygenates and methanol) were observed and characteristic VOC fingerprints of emissions referable to well operations were determined. In general, the separators, dehydrators, pneumatic devices and pumps were identified to emit a mixture of lighter VOC (i.e. higher percentage of compounds with lighter molecular mass), which was similar to the composition of natural gas. Oil and condensate tanks on the other hand were identified to emit heavier VOCs. Downwind of methanol tanks and condensate storage tanks high methanol concentrations were found. Peak mixing ratios of aromatic

hydrocarbons (benzene, toluene, C_8-aromatics, etc.) in the $ppmV$ range were measured at a hydraulically re-fractured gas well with open flow-back pond. These occasionally observed peak values outnumbered typical urban average concentrations for aromatic hydrocarbons by $2-3$ orders of magnitude. The emissions from condensate storage tanks and the open flow-back pond varied in their individual VOC composition, such that the tanks predominantly emitted a mixture of lighter aromatic compounds and the flow-back pond a heavier mixture. Due to favorable wind conditions allowing a differentiation of plume trajectories, emission rates of toluene were estimated using a Lagrangian dispersion model. The comparison of the observations with an emission inventory indicates, that the main emissions in the Uintah Basin were due to individual well pad emissions and thus the observed VOC compositions were mixture of raw natural gas (from the pneumatic devices, pumps and other venting sources) with flashing emissions from storage tanks at oil and gas production sites and from glycol dehydrators. In addition to the characterization of VOC emissions and compositions, hydrogen sulfide (H_2S) was detected at tanks under service.
The presented data were some of the first measurements using fast-response VOC instruments to study emissions at individual well pads and demonstrate the applicability of PTR-MS for this kind of mobile investigations. In this way, the presented measurements and methods provide the basis for further studies, as they filled in the blank between stationary VOC measurements of background mixing ratios and on site characterizations at individual emitters. The results contribute to the understanding of air quality issues in oil and natural gas production areas by identifying of emitters and corresponding VOC fingerprints. A feeding of gathered data into emission inventories will improve future modeling and estimating of air quality and chemical processing in fossil fuel production regions. Moreover, the quantification of VOCs, that act as radical sources (via carbonyl photolysis) within this unusual chemical environment during wintertime conditions, helps to design mitigation strategies for enhanced formation of surface ozone. The complete data record is made available to the scientific community and can be found at the mission home at NOAA ESRL (UBWOS 2012[52]) in the data format ICARTT. Nevertheless, in order to advance our understanding of environmental impacts of fossil fuel production (or energy infrastructures in general), more such measurements are needed. Next steps should thus be:

- distinct enlargement of the data set of ground-based measurements covering the broad variety of VOC emitters in different fossil fuel production areas
- determination of VOC emission rates in addition to ambient mixing ratios (by studying of the vertical extents of individual plumes)
- combination of precise ground-based and large-scale airborne investigations in order to draw conclusions from the individual facility-dependent emitters to the total atmospheric emissions of sectors of industry or energy production in general

52 `http://esrl.noaa.gov/csd/groups/csd7/measurements/2012ubwos/`

The presented instrument ULW-PTR-MS is scheduled to be employed during the upcoming mission Oxidation Mechanisms Observation (OMO), which is a HALO aircraft mission that will take place in 2015 in form of technical demo flights in Europe (January/February) and scheduled measurement flights in Asia in the course of the Indian summer monsoon (July/August). This latter scenario is characterized by heavy vertical mixing across the entire free troposphere and the lower stratosphere and thus suited to allow investigations of free-radical chemistry at higher altitudes. The range of organic trace compounds, that are planned to be measured by ULW-PTR-MS, are expected to contribute to the understanding of the oxidation mechanisms in the atmosphere:

- by determination of the altitude distributions of climate-relevant natural and anthropogenic organic air pollutants with high temporal/spatial resolution
- by identification and quantification of intercontinental long-range transport of NMVOCs at higher altitudes in connection with Brewer-Dobson/Hadley circulations, episodic convective uplift pollution and upward transfer in the warm conveyor belts of cyclonic disturbances
- and by quantification of (photo-chemically active) precursors in the HO_x chemistry, like e.g. acetone, methanol and formaldehyde

Thinking beyond the presented applications, a further deployment of the ULW-PTR-MS aboard mobile laboratories has great potentials to investigate air pollution or other biogenic and anthropogenic processes across many different areas, like e.g. in woodlands, industrial facilities or urban environments.

Appendices

Appendix A

Standard components used in the ULW-PTR-MS

Device assignments according assembly unit descriptions from section 3.1 (p. 37) (if applicable).

Device	Identification	Manufacturer
• **Subrack 19″**	BGTR H M.GR.3HE 84TE 295T KD (incl. accessories)	Schroff
• **AC/DC** $115 - 250\,V\,DC$	VIPAC VP-B1768381	Vicor
• **DC/DC** $5\,V\,DC$	V24C5E50BF	Vicor
• **DC/DC** $-24\,V\,DC$	V24C24C100BL2	Vicor
• **Input filter**	Single-Phase Filter FN 2090	Schaffner
• **Circuit breaker**	4120	E-T-A
• **Circuit breaker**	7274	Klixon
• **Ventilator**	OD6015-24HB	Orion Fans

Table T A.1: *The standard components used for the setup of the* ***Control Unit*** *(see section 3.1.1, p. 40).*

Device	Identification	Manufacturer
• **Subrack 19″**	BGTR H M.GR.3HE 84TE 355T KD (incl. accessories)	Schroff
• **Valves**	225K022/T225PK012	NResearch
• **MFC**$_{BYPASS}$	low-Δp-Flow F-201DV-ABD-22-V 1ln/min	Bronkhorst
• **MFC**$_{H_2O}$	low-Δp-Flow F-200DV-ABD-22-V 15mln/min	Bronkhorst
• **PC**$_{INLET}$	EL-PRESS/ P-702CV-350A-ABD-22-V 150mln/min	Bronkhorst
• **Sensor P**$_{INLET}$	EL-PRESS/ CTE7001AY0	SensorTechnics
• **Sensor P**$_{DT}$	CMR 363	Pfeiffer Vacuum
• **Heating foil**	HK5182R52.9L12B	MINCO
• **CCF**	SUPELPURE HC 2-2445 U	SIGMA-ALDRICH

Table T A.2: *The standard components used for the setup of the* ***Air and Water Flow System*** *(see section 3.1.2, p. 43).*

Device	Identification	Manufacturer
• **TMP**	HiPace80TC110: DN40ISO-KF, DN63ISO-K	Pfeiffer Vacuum

Table T A.3: *The standard components used for the setup of the* ***Intermediate Chambers*** *(see section 3.1.4, p. 49).*

Device	Identification	Manufacturer
• **TMP**	HiPace80TC110: DN40ISO-KF, DN63ISO-K	Pfeiffer Vacuum
• **Sensor P**$_{QUADRUPOLE}$	PBR 260	Pfeiffer Vacuum
• **MS Analyzer**	QMA 400	Pfeiffer Vacuum
• **Ion Counter Preamplifier**	CP 400	Pfeiffer Vacuum
• **SEM**	SEV 217	Pfeiffer Vacuum
• **Ventilators**	SanACE60 109L0624S401	Telemeter

Table T A.4: *The standard components used for the setup of the* ***Quadrupole Mass Spectrometer*** *(see section 3.1.5, p. 50).*

Device	Identification	Manufacturer
• **Diaphragm pump**	MD1VARIO-SP	VACCUBRAND
• **High frequency generator**	QMH 400-5	Pfeiffer Vacuum
• **Wire rope isolators**	WR2-600-10 DM	Enidine

Table T A.5: *The standard components used for the setup of the* ***Assistance Systems*** *(see section 3.1.6, p. 51).*

Device	Identification	Manufacturer
• **Cabling**	-diverse-	Telemeter, HUBER+SUHNER
• **Connectors**	-diverse-	Lemo, Amphenol, WAGO, ITT
• **Tubing** (PFA, PTFE, PEEK)	-diverse-	
• **Fittings, connectors**	-diverse-	EM-Technik, Upchurch Scientific, Swagelok
• **Temperature sensors**	NTC $3k\Omega$	RS
• **PT100 sensor**	W-EYK 6	Heraeus
• **Thermal fuses**	-diverse-	
• **Heating cartridge**	High performance heating cartridge $OD = 6.5\,mm$	Elektrowärme Gabbey
• **Heating cords**	FLEXCORD C1S	flexelec

Table T A.6: *Other standard components.*

Appendix B

Operations/System Control

The ULW-PTR-MS is designed to be used either interactively in the laboratory or unattended during field/aircraft missions. The system architecture provides consecutive system states, that are controlled by the newly developed software being based on the independent V25 operating system. A sequence of operation diagram illustrating the typical process is presented in table T B.1 (p. 164). During manual control the states are individually selectable (e.g. for maintenance, laboratory work). When using the automatic mode, the instrument sequentially passes through the individual system states that follow each other in both ascending and descending order. The stepwise start-up strategy is a concession to a potentially limited peak power availability, like e.g. during battery supply (not applicable for the HALO or CARIBIC aircraft). The system is permanently supervised by a watchdog timer and scenarios, that need restart or guided shutdown of the system - like after a failure of system components, total breakdown of the vacuum or complete consumption of water resources in the H_2O tank - are handled following defined procedures.

▶ **Status: IDLE** (entire system in idle state)

- **Housekeeping**: continuous surveillance of pressures, temperatures and flows
- **Evacuation 1 - MD1**: backing pump MD1 pumps the instrument to internal pressures $< 2\,hPa$; all valves are closed
- **Evacuation 2 - TMP$_{\#1}$**: turbo molecular pump TMP$_{\#1}$ starts evacuating the drift tube and upper part of the intermediate chamber when internal pressure $P_{drift} < 5\,hPa$
- **Evacuation 3 - TMP$_{\#2}$ & TMP$_{\#3}$**: turbo molecular pumps TMP$_{\#2}$ and TMP$_{\#3}$ start evacuating the intermediate and high vacuum chambers when rotation speed of TMP$_{\#1}$ is $\geq 800\,rpm$

▶ **Status: HIGH VACUUM** (chamber pressures ready-to-operate)

- **Flushing 1 - IS**: flushing of ion source with constant flow of saturated water vapor from the reservoir tank ($t_{flush-H_2O} = 60\,sec$)
- **Flushing 2 - Inlet System**: flushing of complete inlet system with ambient air; relevant valves open ($t_{flush-air} = 100\,sec$)
- **Thermal Controls**: heating of DT, water supply pipe, inlet system, scrubbing unit, etc. (continuous PID[a] regulation)

▶ **Status: FLOWS** (inflows of sample air and H_2O vapor ready-to-operate)

- **High Voltages**: supply of IS, DT electrodes and SEM with appropriate high-voltages
- **QMS/RF**: measurements according programmed m/z sequences
- **Internal calibrations**: AMBIENT/BACKGROUND channels controlled according programmed measurement pattern

▶ **Status: FULL OPERATION** (measuring mode)

[a] PID: proportional, integral, differential

Table T B.1: *Sketch of a typical operations procedure of the ULW-PTR-MS depicting the four major system states IDLE, HIGH VACUUM, FLOWS and FULL OPERATION, as well as some minor intermediate states.*

Appendix C

Data processing & Peak Evaluation

Standard post-processing:

The measurement data presented within this work have been post-processed using a set of processing tools that run on the platform IGOR Pro[53]. The used procedures have been oriented on tools used in connection with the data acquisition and management during the mission UBWOS and provide processed final data in the format ICARTT[54].

Peak evaluation:

The analysis engine embedded in the V25 module controlling the QMS allows different ways of mass peak analysis. The data structure provides 8 independent micro-scans (n: number of micro-scan) per examined peak with scanned total width W in $[amu]$:

$$W = \sum_{n=1}^{8} W_{micro-scan(n)} \qquad \text{(E C.1)}$$

W in equation E C.1 (p. 165) is individually selectable:

- $\boxed{W = 1} \rightarrow$ 8 steps over one m/z (e.g. from m/z 36.5 to m/z 37.5)
  $\Rightarrow$ evaluation of peak shape
- $\boxed{W > 1} \rightarrow$ like $W = 1$ but over larger mass range (e.g. m/z 21 to m/z 59)
  $\Rightarrow$ scanning over multiple m/z
- $\boxed{W < 1} \rightarrow$ 8 steps over selected (reduced) range (e.g. from m/z 36.9 to m/z 37.1)
  $\Rightarrow$ enhancement of mass/time resolution, when using multiple points per scan cycle

---

[53] IGOR Pro: numerical computing environment from WaveMetrics (http://www.wavemetrics.com/)

[54] International Consortium for Atmospheric Research on Transport and Transformation (for detailed format descriptions see Aknan *et al.* [2013])

Figure F C.1 (p. 167) gives an example of the mass peak analysis procedure presenting a laboratory scanning of a peak with $W = 1$ around the center mass m/z 37.0 (recorded with integration time $t_{int} = 100\,ms$ per micro-scan; discrimination level: $0.3\,V$). For the peak evaluation different approaches are available, that together allow a multifaceted analysis of the measured peak:

1. **Peak Maximum $\leftrightarrow$ micro-scan with maximum value**
   here: pot #4 with $i(m/z37)_{\#4} \simeq 31.85 \times 10^3\, cps$
2. **Center Mass $\leftrightarrow$ micro-scan including the center mass**
   here: pot #5 with $i(m/z37)_{\#5} \simeq 28.02 \times 10^3\, cps$
3. **Interpolated Curve $\leftrightarrow$ maximum value**
   here: at $\sim$ m/z 36.9 with $i(m/z37)_{max\_interpolated} \simeq 32.63 \times 10^3\, cps$
4. **Integral $\leftrightarrow$ sum of (selected) micro-scans**
   here: for all 8 data pots: $i(m/z37)_{integral(\#1-8)} \simeq 130.24 \times 10^3\, cps$
5. **Gauss-fit $\leftrightarrow$ maximum value**
   here: data pot #4 with $i(m/z37)_{gauss-fit} \simeq 33.60 \times 10^3\, cps$

The full set of available information allows more comprehensive quality checks for the scanned peak. However, during normal measurements typically either of the first two approaches (① Peak Maximum or ② Center Mass) is chosen. This also corresponds to the width reduction of the scan area ($\boxed{W < 1}$) towards concentrated measuring points inside of predetermined peak value regions. Since the observed count rates typically differ between the presented methods (if $W \neq 0$; as shown above), any scientific evaluation of peak data gathered in either of these ways must be adapted to the selected strategy by means of an individual deduction of corresponding calibration parameters.

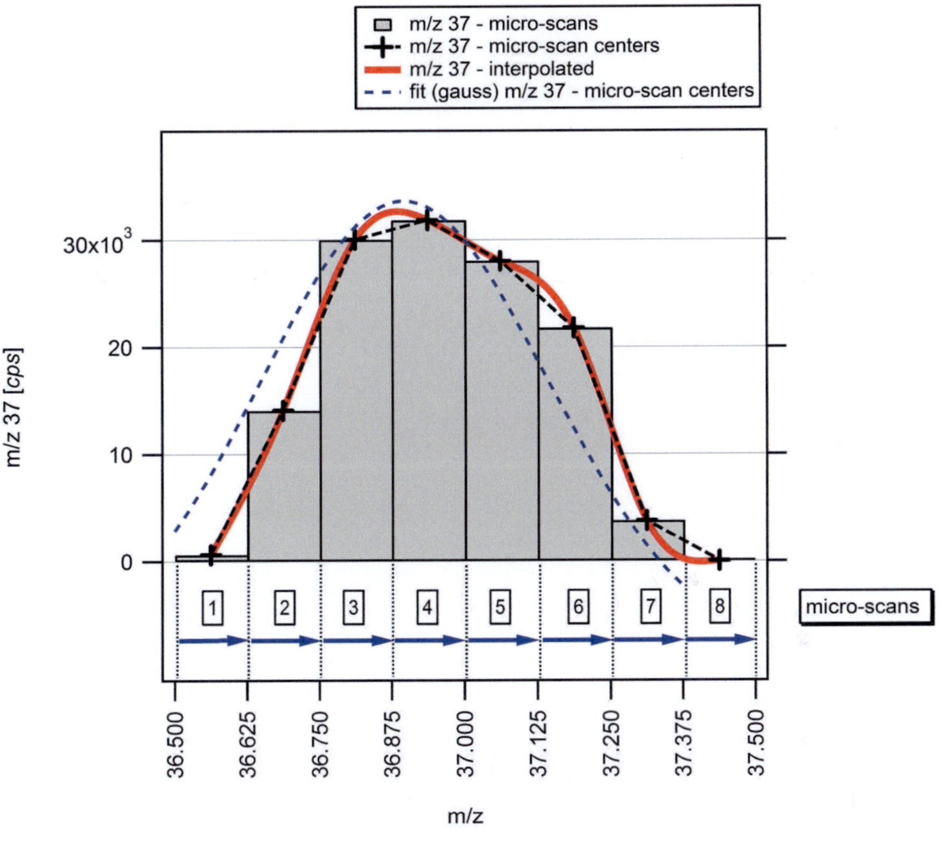

Figure F C.1: *Scanning of m/z 37, that demonstrate the 8 micro-scans over the full width of* $W = 1$ *([amu]) in the range of [36.5; 37.5] showing data bars, the interpolation line and a gauss-fit of the mass peak.*

Appendix D

Flight tests during MUSICA 2013

First-time aircraft measurements in the context of the mission MUSICA 2013 have been performed using the MUSICA PTR-MS, which was the sister instrument of ULW-PTR-MS. Any scientific attendances during these flight tests were beyond the horizon of this work. Nevertheless, as this testing (used primarily for instrumental evaluations) represents an important step within the system development, a brief evaluation of some results is therefore attached to this work for the sake of completeness.

MUSICA 2013 The atmospheric water cycle is an important aspect of the Earth's climate system. During MUSICA[55] (MUlti-platform remote Sensing of Isotopologues for investigating the Cycle of Atmospheric water) this cycle was investigated using different techniques for measurements of H_2O isotopologues: ① satellite-borne and ground based remote sensing instruments and ② airborne and ground based in situ instruments. In order to perform simultaneous validation measurements of instruments determining stable water isotopologues ($H_2^{16}O$, $HD^{16}O$, $H_2^{18}O$ and their fractions δD, $\delta^{18}O$), several research flights over the Canary Islands/Spain have taken place in July 2013 (HQs at Base Aérea de Gando on Gran Canaria/Spain). The involved instruments comprised the ground based FTIR and in situ spectrometers stationed at the Izaña Atmospheric Observatory on Tenerife, Spain, the infrared sensor IASI on the satellite METOP [Schneider & Hase, 2011] and the airborne in situ diode-laser spectrometer ISOWAT [Dyroff *et al.*, 2009].

Carrier CASA C-212 The aircraft instruments have been carried by the twin engine research aircraft CASA C-212 from the Spanish National Institute of Aerospace Technology[56] (INTA), presented in figure F D.1 (p. 170). This modified cargo plane was suitable for the required precomputed aerial maneuvers during ascents and descents allowing the scheduled parallel measurements with the suite of instruments (ground, satellite, aircraft) within the reference area. Besides the core mission instrumentation for the investigation of the water isotopologues and aerosol, the payload of the aircraft facilitated the additional deployment of the PTR mass spectrometer for instrumental testings during the ferry and research flights.

55 `http://www.imk-asf.kit.edu/english/musica.php`

56 `http://www.inta.es/`

Figure F D.1: MUSICA aircraft CASA C-212 during preparation for takeoff at the Torrejón Air Base in Madrid/Spain for the ferry flight to the Naval Station Rota/Spain on 5 July 2013.

MUSICA PTR-MS The MUSICA PTR-MS was a recreation of the ULW-PTR-MS, that was adapted to be deployed to the aircraft CASA C-212 in a project separated from this work. The need for this further instrument had arisen as the ULW-PTR-MS was originally scheduled to be deployed aboard HALO exactly at that time and a comparable instrument was necessary for the planned parallel use during MUSICA 2013. However, the final completion of the MUSICA PTR-MS had not been fully achieved by the begin of the mission in June 2013 and the controls of ULW-PTR-MS, which were available on short call due to another postponement of the HALO mission OMO, were inserted in exchange. In spite of instrumental accordance, this last-minute combination and the general status of the MUSICA PTR-MS, which was untested and uncharacterized at that time, did not allow reasonable atmospheric high-sensitivity measurements.

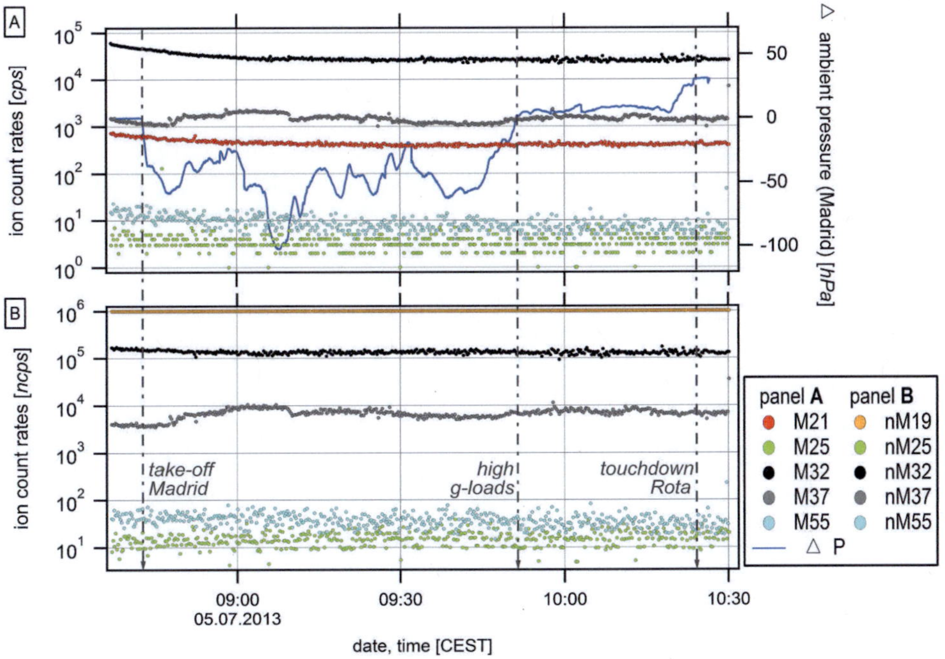

Figure F D.2: *Time series for primary signals $H_3^{16}O^+$ (m/z 19) and $H_3^{18}O^+$ (m/z 21) (constant isotope ratio $^{16}O/_{18}O \sim 500$), instrumental noise (m/z 25), oxygen indicator O_2^+ (m/z 32), water dimer $H_3O^+(H_2O)$ (m/z 37) and water trimer $H_3O^+(H_2O)_2$ (m/z 55) measured during the ferry flight from Madrid/Spain to Rota/Spain during MUSICA 2013 on 5 July 2013 (panel A in [cps]; panel B in [ncps]). Δ ambient pressure (Madrid) in [hPa] and characteristic features of the flight are plotted.*

Demonstration flight attesting the instrument's basic airworthiness Figure F D.2 (p. 171) shows time series of selected signals that illustrate intrasystem processes (IS output) measured with integration times of $100\,ms$ per m/z. The data originated from the two hour ferry flight on 5 July 2013 from INTA HQs in Madrid/Spain to an airfield in Rota/Spain, where the aircraft was refueled for the further transfer to Gran Canaria/Spain. The flight was performed at low altitudes above ground following the orography and comprised an ambient pressure gradient of $\Delta P \simeq 131\,hPa$ (taken from P_{inlet} of ISOWAT). The background noise level on m/z 25 was low during the entire flight with an average rate of $i(M25) \simeq 3.8\,cps$. The only observable external influences onto the shown series were fluctuations in the water dimer $H_3O^+(H_2O)$ signals on m/z 37 indicating normal interferences with ambient air water concentrations. High g-loads during the flight did not affect the system operation, like e.g. during a low level pass over a lake near Cordoba/Spain at 09:50.

The touchdown at Rota/Spain was at 10:22. Besides the mentioned issues with the overall performance, the system itself operated without further interferences during this demonstration flight.

This brief example generally demonstrates the basic airworthiness of the new PTR systems. After the following revision of the MUSICA instrument, the future system will be adapted for its scheduled deployment aboard the CARIBIC aircraft in spring 2015.

Appendix E

HALO Flight Rehearsal for OMO 2015 - Impressions

Figure F E.1 (p. 174) shows some impressions from the HALO aircraft as received during the flight rehearsal for the mission OMO at the aircraft HQ in Oberpfaffenhofen/Germany.

(a) *Front view with noseboom*

(b) *Overview, left wing*

(c) *Different trace gas inlets (TGI) and viewports on top surface*

(d) *Left engine; opening below: baggage compartment aperture*

(e) *Right engine in front of tail fin*

(f) *View into the flight deck*

Figure F E.1: *Impressions from the HALO aircraft during the flight rehearsal inside the aircraft hangar at HQ in Oberpfaffenhofen/Germany.*

Appendix F

UBWOS 2012 - Impressions

Figure F F.1 (p. 176) presents some impressions as received during the field intensive UBWOS 2012 in the Uintah Basin, Utah/USA.

(a) *Uintah Basin gas field with smoke trail originating from a fire at Glen Bench*

(b) *Pump jacks at an oil well in the fields of Duchesne County*

(c) *Water injection well*

(d) *Dried out collection/evaporation reservoir with plastic foil*

(e) *Horse Pool field site with core team*

(f) *Mobile laboratory during sunset*

Figure F.F.1: *Impressions from UBWOS 2012.*

Appendix G

Hydrogen sulfide (H_2S) calibrations

Studies	Time	Instruments	Purposes
UBWOS 2012	Jan - Feb 2012	ULW-PTR-MS	Mobile measurements
UBWOS 2013	Jan - Feb 2013	NOAA PTR-MS PTR-ToF-MS Picarro	Ground measurements at Horse Pool
Laboratory	Mar 2013	NOAA PTR-MS PTR-ToF-MS	Calibrations & Detection humidity dependence

Table T G.1: *List of studies and instruments from which the H_2S data were used for the deduction of the water-sensitive calibration curves and the ambient mixing ratios measured during UBWOS 2012. (adopted from Li* et al. *[2014])*

	IS: Water flow [$sccm$]	DT: Pressure [hPa]	DT: Voltage [V]	DT: Temperature [$°C$]	DT: E/N [Td]
NOAA PTR-MS	10.5	2.4	720	45	117
PTR-ToF-MS	5	2.2	600	60	112
ULW-PTR-MS	7.5	2.2	612	40	106

Table T G.2: *Instrument settings of NOAA PTR-MS, PTR-ToF-MS and ULW-PTR-MS that affect the humidity-dependent sensitivities.*
(adopted from Li et al. *[2014])*

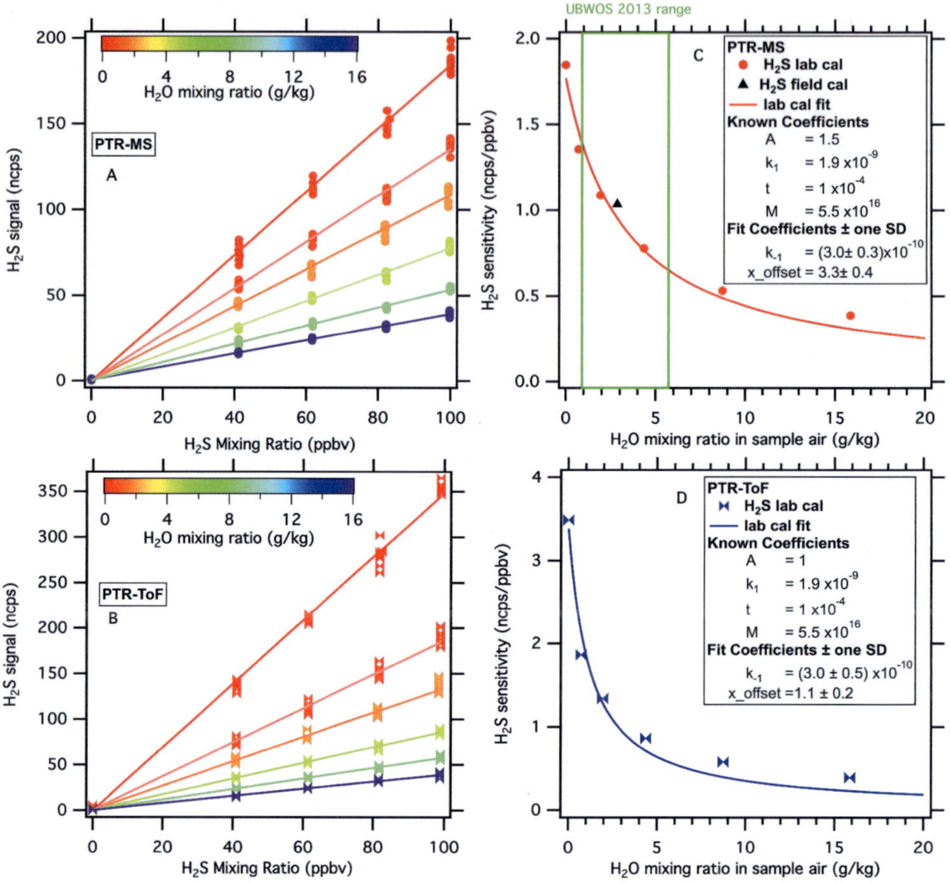

Figure F G.1: *Hydrogen sulfide laboratory calibration curves at different humidities for* A *NOAA PTR-MS (average* H_3O^+ *ion signals of* $25 \times 10^6 cps$*),* B *PTR-ToF-MS (average* H_3O^+ *ion signal of* $8.8 \times 10^6 cps$ *duty cycle corrected and* $1.2 \times 10^6 cps$ *actual counts). The sensitivities for* H_2S *measurements for* C *NOAA PTR-MS and* D *PTR-ToF-MS dependent on humidity.*
(adopted from Li et al. *[2014])*

References

AKNAN, A., CHEN, G., CRAWFORD, J. & WILLIAMS, E. (2013). International Consortium for Atmospheric Research on Transport and Transformation (ICARTT) File Format Standards V1.1. Tech. rep., National Aeronautics and Space Administration (NASA). 165

ALLEN, D.T., TORRES, V.M., THOMAS, J., SULLIVAN, D.W., HARRISON, M., HENDLER, A., HERNDON, S.C., KOLB, C.E., FRASER, M.P., HILL, A.D., LAMB, B.K., MISKIMINS, J., SAWYER, R.F. & SEINFELD, J.H. (2013). Measurements of methane emissions at natural gas production sites in the United States. *Proceedings of the National Academy of Sciences of the United States of America*, **110**, 17768–73. 102

ANDREAE, M.O., ARTAXO, P., FISCHER, H., FREITAS, S.R., GRÉGOIRE, J.M., HANSEL, A., HOOR, P., KORMANN, R., KREJCI, R., LANGE, L., LELIEVELD, J., LINDINGER, W., LONGO, K., PETERS, W., DE REUS, M., SCHEEREN, B., SILVA DIAS, M.A.F., STRÖM, J., VAN VELTHOVEN, P.F.J. & WILLIAMS, J. (2001). Transport of biomass burning smoke to the upper troposphere by deep convection in the equatorial region. *Geophysical Research Letters*, **28**, 951–954. 84

APEL, E.C., BRAUERS, T., KOPPMANN, R., BANDOWE, B., BOSSMEYER, J., HOLZKE, C., TILLMANN, R., WAHNER, A., WEGENER, R., BRUNNER, A., JOCHER, M., RUUSKANEN, T., SPIRIG, C., STEIGNER, D., STEINBRECHER, R., GOMEZ ALVAREZ, E., MÜLLER, K., BURROWS, J.P., SCHADE, G., SOLOMON, S.J., LADSTÄTTER-WEISSENMAYER, A., SIMMONDS, P., YOUNG, D., HOPKINS, J.R., LEWIS, A.C., LEGREID, G., REIMANN, S., HANSEL, A., WISTHALER, A., BLAKE, R.S., ELLIS, A.M., MONKS, P.S. & WYCHE, K.P. (2008). Intercomparison of oxygenated volatile organic compound measurements at the SAPHIR atmosphere simulation chamber. *Journal of Geophysical Research*, **113**, 1–24. 62

ARNOLD, S.R., CHIPPERFIELD, M.P., BLITZ, M.A., HEARD, D.E. & PILLING, M.J. (2004). Photodissociation of acetone: Atmospheric implications of temperature-dependent quantum yields. *Geophysical Research Letters*, **31**, 1–4. 4

ASCHE, F., OGLEND, A. & OSMUNDSEN, P. (2012). Gas versus oil prices the impact of shale gas. *Energy Policy*, **47**, 117–124. 100

ATKINSON, R. (2000). Atmospheric chemistry of VOCs and NOx. *Atmospheric Environment*, **34**, 2063–2101. 4

REFERENCES

ATKINSON, R. (2007). Gas-phase tropospheric chemistry of organic compounds: a review. *Atmospheric Environment*, **41**, 200–240. 4

ATKINSON, R. & AREY, J. (2003a). Atmospheric degradation of volatile organic compounds. *Chemical Reviews*, **103**, 4605–38. 4, 5, 9

ATKINSON, R. & AREY, J. (2003b). Gas-phase tropospheric chemistry of biogenic volatile organic compounds: a review. *Atmospheric Environment*, **37**, 197–219. 8

ATSDR (2000). Toxicological profile for toluene. Tech. Rep. September, US Department of Health and Human Services, Public Health Service, Agency for Toxic Substances and Disease Registry (ATSDR). 99

ATSDR (2006). Toxicological profile for hydrogen sulfide. Tech. rep., US Department of Health and Human Services, Public Health Service, Agency for Toxic Substances and Disease Registry (ATSDR). 149, 151

ATSDR (2007). Toxicological profile for benzene. Tech. Rep. August, US Department of Health and Human Services, Public Health Service, Agency for Toxic Substances and Disease Registry (ATSDR). 98

BAHREINI, R., MIDDLEBROOK, A.M., BROCK, C.A., DE GOUW, J.A., MCKEEN, S.A., WILLIAMS, L.R., DAUMIT, K.E., LAMBE, A.T., MASSOLI, P., CANAGARATNA, M.R., AHMADOV, R., CARRASQUILLO, A.J., CROSS, E.S., ERVENS, B., HOLLOWAY, J.S., HUNTER, J.F., ONASCH, T.B., POLLACK, I.B., ROBERTS, J.M., RYERSON, T.B., WARNEKE, C., DAVIDOVITS, P., WORSNOP, D.R. & KROLL, J.H. (2012). Mass spectral analysis of organic aerosol formed downwind of the Deepwater Horizon oil spill: field studies and laboratory confirmations. *Environmental Science & Technology*, **46**, 8025–34. 10

BAKER, A.K., TRAUD, S., BRENNINKMEIJER, C.A., HOOR, P., NEUMAIER, M., ORAM, D.E., RAUTHE-SCHÖCH, A., SPRUNG, D., SCHLOEGL, S., SLEMR, F., VAN VELTHOVEN, P.F., WERNLI, H., ZAHN, A. & ZIEREIS, H. (2014). Pollution patterns in the upper troposphere over Europe and Asia observed by CARIBIC. *Atmospheric Environment*, **96**, 245–256. 84

BHASKARAN, B., JONES, R.G., MURPHY, J.M. & NOGUER, M. (1996). Simulations of the Indian summer monsoon using a nested regional climate model: domain size experiments. *Climate Dynamics*, **12**, 573–587. 88

BLAKE, R.S., MONKS, P.S. & ELLIS, A.M. (2009). Proton-transfer reaction mass spectrometry. *Chemical Reviews*, **109**, 861–96. 11, 17, 23, 49, 58

BORBON, A., GILMAN, J.B., KUSTER, W.C., GRAND, N., CHEVAILLIER, S., COLOMB, A., DOLGOROUKY, C., GROS, V., LOPEZ, M., SARDA-ESTEVE, R., HOLLOWAY, J., STUTZ, J., PETETIN, H., MCKEEN, S., BEEKMANN, M.,

WARNEKE, C., PARRISH, D.D. & DE GOUW, J.A. (2013). Emission ratios of anthropogenic volatile organic compounds in northern mid-latitude megacities: Observations versus emission inventories in Los Angeles and Paris. *Journal of Geophysical Research: Atmospheres*, **118**, 2041–2057. 10, 116

BRASSEUR, G. & SOLOMON, S. (1986). *Aeronomy of the Middle Atmosphere*, vol. 32 of *Atmospheric and Oceanographic Sciences Library*. Springer Netherlands, Dordrecht. 3

BRENNINKMEIJER, C.A.M., CRUTZEN, P.J., FISCHER, H., GÜSTEN, H., HANS, W., HEINRICH, G., HEINTZENBERG, J., HERMANN, M., IMMELMANN, T., KERSTING, D., MAISS, M., NOLLE, M., PITSCHEIDER, A., POHLKAMP, H., SCHARFFE, D., SPECHT, K. & WIEDENSOHLER, A. (1999). CARIBICCivil Aircraft for Global Measurement of Trace Gases and Aerosols in the Tropopause Region. *Journal of Atmospheric and Oceanic Technology*, **16**, 1373–1383. 35, 83

BRENNINKMEIJER, C.A.M., CRUTZEN, P., BOUMARD, F., DAUER, T., DIX, B., EBINGHAUS, R., FILIPPI, D., FISCHER, H., FRANKE, H., FRIEβ, U., HEINTZENBERG, J., HELLEIS, F., HERMANN, M., KOCK, H.H., KOEPPEL, C., LELIEVELD, J., LEUENBERGER, M., MARTINSSON, B.G., MIEMCZYK, S., MORET, H.P., NGUYEN, H.N., NYFELER, P., ORAM, D., O'SULLIVAN, D., PENKETT, S., PLATT, U., PUPEK, M., RAMONET, M., RANDA, B., REICHELT, M., RHEE, T.S., ROHWER, J., ROSENFELD, K., SCHARFFE, D., SCHLAGER, H., SCHUMANN, U., SLEMR, F., SPRUNG, D., STOCK, P., THALER, R., VALENTINO, F., VAN VELTHOVEN, P., WAIBEL, A., WANDEL, A., WASCHITSCHEK, K., WIEDENSOHLER, A., XUEREF-REMY, I., ZAHN, A., ZECH, U. & ZIEREIS, H. (2007). Civil Aircraft for the regular investigation of the atmosphere based on an instrumented container: The new CARIBIC system. *Atmospheric Chemistry and Physics*, **7**, 4953–4976. 83

BRITO, J. & ZAHN, A. (2011). An unheated permeation device for calibrating atmospheric VOC measurements. *Atmospheric Measurement Techniques*, **4**, 2143–2152. 52

BRITO, J.F.D. (2011). *A Lightweight High-sensitivity Chemical Mass Spectrometer for Organic Compounds*. Ph.D. thesis, Karlsruhe Institute of Technology (KIT). 35, 50, 55

CADET, D. (1979). Meteorology of the Indian summer monsoon. *Nature*, **279**, 761–767. 88

CAMPANA, J. (1980). Elementary theory of the quadrupole mass filter. *International Journal of Mass Spectrometry and Ion Physics*, **33**, 101–117. 30

CARBAJO, P.G., SMITH, S.C., HOLLOWAY, A.L., SMITH, C.A., POPE, F.D., SHALLCROSS, D.E. & ORR-EWING, A.J. (2008). Ultraviolet photolysis of HCHO: absolute HCO quantum yields by direct detection of the HCO radical photoproduct. *The Journal of Physical Chemistry A*, **112**, 12437–48. 115

CARLTON, A.G., LITTLE, E., MOELLER, M., ODOYO, S. & SHEPSON, P.B. (2014). The data gap: can a lack of monitors obscure loss of clean air act benefits in fracking areas? *Environmental Science & Technology*, **48**, 893–4. 100

CARTER, W.P.L. & SEINFELD, J.H. (2012). Winter ozone formation and VOC incremental reactivities in the Upper Green River Basin of Wyoming. *Atmospheric Environment*, **50**, 255–266. 8, 100

CHIDSEY JR., T.C. & LAINE, M.D. (1992). The fractured Green River and Wasatch formations of the Uinta basin, Utah: targets for horizontal drilling. *Hydrocarbon and Mineral Resources of the Uinta Basin, Utah and Colorado*, 123–134. 99

CLAEYS, M., GRAHAM, B., VAS, G., WANG, W., VERMEYLEN, R., PASHYNSKA, V., CAFMEYER, J., GUYON, P., ANDREAE, M.O., ARTAXO, P. & MAENHAUT, W. (2004). Formation of secondary organic aerosols through photooxidation of isoprene. *Science (New York, N.Y.)*, **303**, 1173–6. 9

COLBORN, T., KWIATKOWSKI, C., SCHULTZ, K. & BACHRAN, M. (2011). Natural Gas Operations from a Public Health Perspective. *Human and Ecological Risk Assessment: An International Journal*, **17**, 1039–1056. 98, 100

CRENNA, B., FLESCH, T. & WILSON, J. (2008). Influence of sourcesensor geometry on multi-source emission rate estimates. *Atmospheric Environment*, **42**, 7373–7383. 146

CROSSON, E. (2008). A cavity ring-down analyzer for measuring atmospheric levels of methane, carbon dioxide, and water vapor. *Applied Physics B*, **92**, 403–408. 108

CROUNSE, J.D., KNAP, H.C., ORNSO, K.B., JORGENSEN, S., PAULOT, F., KJAERGAARD, H.G. & WENNBERG, P.O. (2012). Atmospheric fate of methacrolein. 1. Peroxy radical isomerization following addition of OH and O2. *The Journal of Physical Chemistry A*, **116**, 5756–62. 9

CRUTZEN, P. (1974). Photochemical reactions initiated by and influencing ozone in unpolluted tropospheric air. *Tellus*, **80302**. 8

CURRIE, L.A. (1968). Limits for qualitative detection and quantitative determination. Application to radiochemistry. *Analytical Chemistry*, **40**, 586–593. 24

DALLAS, A.J. & CARR, P.W. (1991). Direct chromatographic comparison of the relative adsorption activity of various types of capillary transfer tubing. *Analytica Chimica Acta*, **251**, 83–93. 44

DAWSON, P.H., ed. (1976). *Quadrupole Mass Spectrometry and Its Applications*. 11, 29, 30, 31, 50

DE GOUW, J. & JIMENEZ, J.L. (2009). Organic aerosols in the Earth's atmosphere. *Environmental Science & Technology*, **43**, 7614–8. 10

DE GOUW, J. & WARNEKE, C. (2007). Measurements of volatile organic compounds in the earth's atmosphere using proton-transfer-reaction mass spectrometry. *Mass Spectrometry Reviews*, **26**, 223–57. 5, 11, 17, 23, 24, 25, 47, 57, 58, 62, 63, 64, 72, 75, 108

DE GOUW, J., WARNEKE, C., KARL, T., EERDEKENS, G., VAN DER VEEN, C. & FALL, R. (2003). Sensitivity and specificity of atmospheric trace gas detection by proton-transfer-reaction mass spectrometry. *International Journal of Mass Spectrometry*, **223-224**, 365–382. 20, 23

DE GOUW, J.A. (2003). Validation of proton transfer reaction-mass spectrometry (PTR-MS) measurements of gas-phase organic compounds in the atmosphere during the New England Air Quality Study (NEAQS) in 2002. *Journal of Geophysical Research*, **108**, 4682. 23, 25, 58, 73

DE GOUW, J.A. (2005). Budget of organic carbon in a polluted atmosphere: Results from the New England Air Quality Study in 2002. *Journal of Geophysical Research*, **110**, D16305. 5, 116

DE GOUW, J.A., MIDDLEBROOK, A.M., WARNEKE, C., AHMADOV, R., ATLAS, E.L., BAHREINI, R., BLAKE, D.R., BROCK, C.A., BRIOUDE, J., FAHEY, D.W., FEHSENFELD, F.C., HOLLOWAY, J.S., LE HENAFF, M., LUEB, R.A., MCKEEN, S.A., MEAGHER, J.F., MURPHY, D.M., PARIS, C., PARRISH, D.D., PERRING, A.E., POLLACK, I.B., RAVISHANKARA, A.R., ROBINSON, A.L., RYERSON, T.B., SCHWARZ, J.P., SPACKMAN, J.R., SRINIVASAN, A. & WATTS, L.A. (2011). Organic aerosol formation downwind from the Deepwater Horizon oil spill. *Science (New York, N.Y.)*, **331**, 1295–9. 10, 146

DERWENT, R. (1995). Sources, distributions, and fates of VOCs in the atmosphere. *Issues in Environmental Science and Technology*, **4**, 1–16. 4

DLR (2009). HALO - Technical Notes. Tech. rep., German Aerospace Center (DLR), Flight Operations (FX). 37, 40, 43

DOTAN, I., ALBRITTON, D.L., LINDINGER, W. & PAHL, M. (1976). Mobilities of CO2+, N2H+, H3O+, H3O+H2O, and H3O+ (H2O)2 ions in N2. *The Journal of Chemical Physics*, **65**, 5028. 20

DOUGLAS, D.J. (2009). Linear quadrupoles in mass spectrometry. *Mass Spectrometry Reviews*, **28**, 937–60. 11, 26, 28, 30

DUNNE, E., GALBALLY, I.E., LAWSON, S. & PATTI, A. (2012). Interference in the PTR-MS measurement of acetonitrile at m/z 42 in polluted urban airA study using switchable reagent ion PTR-MS. *International Journal of Mass Spectrometry*, **319-320**, 40–47. 84

DYROFF, C., FÜTTERER, D. & ZAHN, A. (2009). Compact diode-laser spectrometer ISOWAT for highly sensitive airborne measurements of water-isotope ratios. *Applied Physics B*, **98**, 537–548. 169

EDTBAUER, A., HARTUNGEN, E., JORDAN, A., HANEL, G., HERBIG, J., JÜRSCHIK, S., LANZA, M., BREIEV, K., MÄRK, L. & SULZER, P. (2014). Theory and practical examples of the quantification of CH4, CO, O2, and CO2 with an advanced proton-transfer-reaction/selective-reagent-ionization instrument (PTR/SRI-MS). *International Journal of Mass Spectrometry*, **365-366**, 10–14. 12

EDWARDS, P.M., YOUNG, C.J., AIKIN, K., DEGOUW, J., DUBÉ, W.P., GEIGER, F., GILMAN, J., HELMIG, D., HOLLOWAY, J.S., KERCHER, J., LERNER, B., MARTIN, R., MCLAREN, R., PARRISH, D.D., PEISCHL, J., ROBERTS, J.M., RYERSON, T.B., THORNTON, J., WARNEKE, C., WILLIAMS, E.J. & BROWN, S.S. (2013). Ozone photochemistry in an oil and natural gas extraction region during winter: simulations of a snow-free season in the Uintah Basin, Utah. *Atmospheric Chemistry and Physics*, **13**, 8955–8971. 8, 113, 114

EDWARDS, P.M., BROWN, S.S., ROBERTS, J.M., AHMADOV, R., BANTA, R.M., DEGOUW, J.A., DUBÉ, W.P., FIELD, R.A., FLYNN, J.H., GILMAN, J.B., GRAUS, M., HELMIG, D., KOSS, A., LANGFORD, A.O., LEFER, B.L., LERNER, B.M., LI, R., LI, S.M., MCKEEN, S.A., MURPHY, S.M., PARRISH, D.D., SENFF, C.J., SOLTIS, J., STUTZ, J., SWEENEY, C., THOMPSON, C.R., TRAINER, M.K., TSAI, C., VERES, P.R., WASHENFELDER, R.A., WARNEKE, C., WILD, R.J., YOUNG, C.J., YUAN, B. & ZAMORA, R. (2014). High winter ozone pollution from carbonyl photolysis in an oil and gas basin. *Nature*. 8, 113, 114, 115

EIA (2013). Annual Energy Outlook 2013 Early Release Overview. Tech. rep., United States Energy Information Administration (EIA). 97, 98

EIA (2014). Annual Energy Outlook 2014 Early Release Overview. Tech. rep., United States Energy Information Administration (EIA). 97

ELIAS, T., SZOPA, S., ZAHN, A., SCHUCK, T., BRENNINKMEIJER, C., SPRUNG, D. & SLEMR, F. (2011). Acetone variability in the upper troposphere: analysis of CARIBIC observations and LMDz-INCA chemistry-climate model simulations. *Atmospheric Chemistry and Physics*, **11**, 8053–8074. 84

ELLIS, A.M. & MAYHEW, C.A. (2014). *Proton Transfer Reaction Mass Spectrometry*. John Wiley & Sons, Ltd, Chichester, UK. 16, 18, 20, 24

ELLIS, H., MCDANIEL, E., ALBRITTON, D., VIEHLAND, L., LIN, S. & MASON, E. (1978). Transport properties of gaseous ions over a wide energy range. Part II. *Atomic Data and Nuclear Data Tables*, **22**, 179–217. 20

ELLIS, H., THACKSTON, M., MCDANIEL, E. & MASON, E. (1984). Transport properties of gaseous ions over a wide energy range. Part III. *Atomic Data and Nuclear Data Tables*, **31**, 113–151. 20

ELLIS, H.H., PAI, R.R., MCDANIEL, E.E., MASON, E. & VIEHLAND, L. (1976). Transport properties of gaseous ions over a wide energy range. *Atomic Data and Nuclear Data Tables*, **17**, 177–210. 20

ELLSWORTH, W.L. (2013). Injection-induced earthquakes. *Science (New York, N.Y.)*, **341**, 1225942. 100

EPA (1993). Report to Congress on hydrogen sulfide air emissions associated with the extraction of oil and natural gas. Final report. Tech. rep., United States Environmental Protection Agency (EPA), Office of Air Quality Planning and Standards. 149

EPA (2012). Our Nation's Air - Status and Trends through 2010. Tech. rep., United States Environmental Protection Agency (EPA). 4, 8, 98

ESSWEIN, E.J., SNAWDER, J., KING, B., BREITENSTEIN, M., ALEXANDER-SCOTT, M. & KIEFER, M. (2014). Evaluation of some potential chemical exposure risks during flowback operations in unconventional oil and gas extraction: preliminary results. *Journal of Occupational and Environmental Hygiene*, **11**, D174–84. 99, 100, 102

FEILBERG, A., LIU, D., ADAMSEN, A.P.S., HANSEN, M.J. & JONASSEN, K.E.N. (2010). Odorant emissions from intensive pig production measured by online proton-transfer-reaction mass spectrometry. *Environmental Science & Technology*, **44**, 5894–900. 149

FIORE, A.M., NAIK, V., SPRACKLEN, D.V., STEINER, A., UNGER, N., PRATHER, M., BERGMANN, D., CAMERON-SMITH, P.J., CIONNI, I., COLLINS, W.J., DALSOREN, S., EYRING, V., FOLBERTH, G.A., GINOUX, P., HOROWITZ, L.W., JOSSE, B., LAMARQUE, J.F., MACKENZIE, I.A., NAGASHIMA, T., O'CONNOR, F.M., RIGHI, M., RUMBOLD, S.T., SHINDELL, D.T., SKEIE, R.B., SUDO, K., SZOPA, S., TAKEMURA, T. & ZENG, G. (2012). Global air quality and climate. *Chemical Society Reviews*, **41**, 6663–83. 6, 10

FLESCH, T., WILSON, J., HARPER, L. & CRENNA, B. (2005). Estimating gas emissions from a farm with an inverse-dispersion technique. *Atmospheric Environment*, **39**, 4863–4874. 146

FLESCH, T.K., WILSON, J.D. & YEE, E. (1995). Backward-Time Lagrangian Stochastic Dispersion Models and Their Application to Estimate Gaseous Emissions. *Journal of Applied Meteorology*, **34**, 1320–1332. 146

FOLKINS, I. & CHATFIELD, R. (2000). Impact of acetone on ozone production and OH in the upper troposphere at high NOx. *Journal of Geophysical Research*, **105**, 11585. 8

FRIED, A., WERT, B.P., HENRY, B. & DRUMMOND, J.R. (1999). Airborne tunable diode laser measurements of formaldehyde. *Spectrochimica Acta Part A: Molecular and Biomolecular Spectroscopy*, **55**, 2097–2110. 68

FUCHS, H., DUBÉ, W.P., LERNER, B.M., WAGNER, N.L., WILLIAMS, E.J. & BROWN, S.S. (2009). A sensitive and versatile detector for atmospheric NO2 and NOx based on blue diode laser cavity ring-down spectroscopy. *Environmental Science & Technology*, **43**, 7831–6. 108

FUCHS, H., HOFZUMAHAUS, A., ROHRER, F., BOHN, B., BRAUERS, T., DORN, H.P., HÄSELER, R., HOLLAND, F., KAMINSKI, M., LI, X., LU, K., NEHR, S., TILLMANN, R., WEGENER, R. & WAHNER, A. (2013). Experimental evidence for efficient hydroxyl radical regeneration in isoprene oxidation. *Nature Geoscience*, **6**, 1023–1026. 9

GETTELMAN, A., HOOR, P., PAN, L.L., RANDEL, W.J., HEGGLIN, M.I. & BIRNER, T. (2011). The extratropical upper troposphere and lower stratosphere. *Reviews of Geophysics*, **49**, RG3003. 84, 88

GILMAN, J.B., KUSTER, W.C., GOLDAN, P.D., HERNDON, S.C., ZAHNISER, M.S., TUCKER, S.C., BREWER, W.A., LERNER, B.M., WILLIAMS, E.J., HARLEY, R.A., FEHSENFELD, F.C., WARNEKE, C. & DE GOUW, J.A. (2009). Measurements of volatile organic compounds during the 2006 TexAQS/GoMACCS campaign: Industrial influences, regional characteristics, and diurnal dependencies of the OH reactivity. *Journal of Geophysical Research*, **114**, D00F06. 57

GILMAN, J.B., BURKHART, J.F., LERNER, B.M., WILLIAMS, E.J., KUSTER, W.C., GOLDAN, P.D., MURPHY, P.C., WARNEKE, C., FOWLER, C., MONTZKA, S.A., MILLER, B.R., MILLER, L., OLTMANS, S.J., RYERSON, T.B., COOPER, O.R., STOHL, A. & DE GOUW, J.A. (2010). Ozone variability and halogen oxidation within the Arctic and sub-Arctic springtime boundary layer. *Atmospheric Chemistry and Physics*, **10**, 10223–10236. 75

GILMAN, J.B., LERNER, B.M., KUSTER, W.C. & DE GOUW, J.A. (2013). Source Signature of Volatile Organic Compounds from Oil and Natural Gas Operations in Northeastern Colorado. *Environmental Science & Technology*, **47**, 1297–1305. 63, 98, 116

GOLDAN, P.D. (2004). Nonmethane hydrocarbon and oxy hydrocarbon measurements during the 2002 New England Air Quality Study. *Journal of Geophysical Research*, **109**, D21309. 57, 108

GOLDSTEIN, A.H. & GALBALLY, I.E. (2007). Known and Unexplored Organic Constituents in the Earth's Atmosphere. *Environmental Science & Technology*, **41**, 1514–1521. 4

GOOD, A. (1970). Mechanism and Rate Constants of IonMolecule Reactions Leading to Formation of H+(H2O)n in Moist Oxygen and Air. *The Journal of Chemical Physics*, **52**, 222. 72

GRAUS, M., MÜLLER, M. & HANSEL, A. (2010). High resolution PTR-TOF: quantification and formula confirmation of VOC in real time. *Journal of the American Society for Mass Spectrometry*, **21**, 1037–44. 12, 72

HANSEL, A. & WISTHALER, A. (2000). A method for realtime detection of PAN, PPN and MPAN in ambient air. *Geophysical Research Letters*, **27**, 895. 19

HANSEL, A., JORDAN, A., HOLZINGER, R., PRAZELLER, P., VOGEL, W. & LINDINGER, W. (1995). Proton transfer reaction mass spectrometry: on-line trace gas analysis at the ppb level. *International Journal of Mass Spectrometry and Ion Processes*, **149-150**, 609–619. 11, 15, 16

HANSEL, A., SINGER, W., WISTHALER, A., SCHWARZMANN, M. & LINDINGER, W. (1997). Energy dependencies of the proton transfer reactions. *International Journal of Mass Spectrometry and Ion Processes*, **167-168**, 697–703. 17, 65

HARTUNGEN, E. (2005). *Proton Transfer Reaction Mass Spectrometer Improvements And Applications*. Ph.D. thesis, University of Innsbruck. 16, 47

HAYWARD, S., HEWITT, C.N., SARTIN, J.H. & OWEN, S.M. (2002). Performance Characteristics and Applications of a Proton Transfer Reaction-Mass Spectrometer for Measuring Volatile Organic Compounds in Ambient Air. *Environmental Science & Technology*, **36**, 1554–1560. 24

HEIMANN, M. (2010). The carbon cycle in the Earth System. *Research Perspectives of the Max Planck Society 2010+*, 1–11. 4

HELMIG, D., THOMPSON, C.R., EVANS, J., BOYLAN, P., HUEBER, J. & PARK, J.H. (2014). Highly elevated atmospheric levels of volatile organic compounds in the Uintah Basin, Utah. *Environmental Science & Technology*, **48**, 4707–15. 99, 106, 115, 117

HEWITT, C.N., HAYWARD, S. & TANI, A. (2003). The application of proton transfer reaction-mass spectrometry (PTR-MS) to the monitoring and analysis of volatile organic compounds in the atmosphere. *Journal of Environmental Monitoring*, **5**, 1–7. 11

HOLLAS, J.M. (2009). Modern Spectroscopy. 6

HOWARTH, R.W., SANTORO, R. & INGRAFFEA, A. (2011). Methane and the greenhouse-gas footprint of natural gas from shale formations. *Climatic Change*, **106**, 679–690. 98

HOWORKA, F., LINDINGER, W. & PAHL, M. (1973). Ion sampling from the negative glow plasma in a cylindrical hollow cathode. *International Journal of Mass Spectrometry and Ion Physics*, **12**, 67–77. 47

HUNTER, E.P.L. & LIAS, S.G. (1998). Evaluated Gas Phase Basicities and Proton Affinities of Molecules: An Update. *Journal of Physical and Chemical Reference Data*, **27**, 413. 17, 18

INOMATA, S., TANIMOTO, H., KAMEYAMA, S., TSUNOGAI, U., IRIE, H., KANAYA, Y. & WANG, Z. (2008). Technical Note: Determination of formaldehyde mixing ratios in air with PTR-MS: laboratory experiments and field measurements. *Atmospheric Chemistry and Physics*, **8**, 273–284. 65

JACOB, D. (2000). Heterogeneous chemistry and tropospheric ozone. *Atmospheric Environment*, **34**, 2131–2159. 8

JENNER, S. & LAMADRID, A.J. (2013). Shale gas vs. coal: Policy implications from environmental impact comparisons of shale gas, conventional gas, and coal on air, water, and land in the United States. *Energy Policy*, **53**, 442–453. 100

JIMENEZ, J.L., CANAGARATNA, M.R., DONAHUE, N.M., PREVOT, A.S.H., ZHANG, Q., KROLL, J.H., DECARLO, P.F., ALLAN, J.D., COE, H., NG, N.L., AIKEN, A.C., DOCHERTY, K.S., ULBRICH, I.M., GRIESHOP, A.P., ROBINSON, A.L., DUPLISSY, J., SMITH, J.D., WILSON, K.R., LANZ, V.A., HUEGLIN, C., SUN, Y.L., TIAN, J., LAAKSONEN, A., RAATIKAINEN, T., RAUTIAINEN, J., VAATTOVAARA, P., EHN, M., KULMALA, M., TOMLINSON, J.M., COLLINS, D.R., CUBISON, M.J., DUNLEA, E.J., HUFFMAN, J.A., ONASCH, T.B., ALFARRA, M.R., WILLIAMS, P.I., BOWER, K., KONDO, Y., SCHNEIDER, J., DREWNICK, F., BORRMANN, S., WEIMER, S., DEMERJIAN, K., SALCEDO, D., COTTRELL, L., GRIFFIN, R., TAKAMI, A., MIYOSHI, T., HATAKEYAMA, S., SHIMONO, A., SUN, J.Y., ZHANG, Y.M., DZEPINA, K., KIMMEL, J.R., SUEPER, D., JAYNE, J.T., HERNDON, S.C., TRIMBORN, A.M., WILLIAMS, L.R., WOOD, E.C., MIDDLEBROOK, A.M., KOLB, C.E., BALTENSPERGER, U. & WORSNOP, D.R. (2009). Evolution of organic aerosols in the atmosphere. *Science (New York, N.Y.)*, **326**, 1525–9. 9

JOBSON, B.T. & MCCOSKEY, J.K. (2010). Sample drying to improve HCHO measurements by PTR-MS instruments: laboratory and field measurements. *Atmospheric Chemistry and Physics*, **10**, 1821–1835. 68

JORDAN, A., HAIDACHER, S., HANEL, G., HARTUNGEN, E., HERBIG, J., MÄRK, L., SCHOTTKOWSKY, R., SEEHAUSER, H., SULZER, P. & MÄRK, T. (2009a). An online ultra-high sensitivity Proton-transfer-reaction mass-spectrometer combined with switchable reagent ion capability (PTR+SRIMS). *International Journal of Mass Spectrometry*, **286**, 32–38. 12, 47

JORDAN, A., HAIDACHER, S., HANEL, G., HARTUNGEN, E., MÄRK, L., SEEHAUSER, H., SCHOTTKOWSKY, R., SULZER, P. & MÄRK, T. (2009b). A high resolution and high sensitivity proton-transfer-reaction time-of-flight mass spectrometer (PTR-TOF-MS). *International Journal of Mass Spectrometry*, **286**, 122–128. 12

KARION, A., SWEENEY, C., PÉTRON, G., FROST, G., MICHAEL HARDESTY, R., KOFLER, J., MILLER, B.R., NEWBERGER, T., WOLTER, S., BANTA, R., BREWER, A., DLUGOKENCKY, E., LANG, P., MONTZKA, S.A., SCHNELL, R., TANS, P., TRAINER, M., ZAMORA, R. & CONLEY, S. (2013). Methane emissions estimate from airborne measurements over a western United States natural gas field. *Geophysical Research Letters*, **40**, 4393–4397. 98, 115

KARL, T. (2003). Use of proton-transfer-reaction mass spectrometry to characterize volatile organic compound sources at the La Porte super site during the Texas Air Quality Study 2000. *Journal of Geophysical Research*, **108**, 4508. 58

KARL, T., APEL, E., HODZIC, A., RIEMER, D.D., BLAKE, D.R. & WIEDINMYER, C. (2009). Emissions of volatile organic compounds inferred from airborne flux measurements over a megacity. *Atmospheric Chemistry and Physics*, **9**, 271–285. 116

KATZENSTEIN, A.S., DOEZEMA, L.A., SIMPSON, I.J., BLAKE, D.R. & ROWLAND, F.S. (2003). Extensive regional atmospheric hydrocarbon pollution in the southwestern United States. *Proceedings of the National Academy of Sciences of the United States of America*, **100**, 11975–11979. 98

KELLY, R.T., TOLMACHEV, A.V., PAGE, J.S., TANG, K. & SMITH, R.D. (2010). The ion funnel: theory, implementations, and applications. *Mass Spectrometry Reviews*, **29**, 294–312. 55

KEMBALL-COOK, S., BAR-ILAN, A., GRANT, J., PARKER, L., JUNG, J., SANTAMARIA, W., MATHEWS, J. & YARWOOD, G. (2010). Ozone impacts of natural gas development in the Haynesville Shale. *Environmental Science & Technology*, **44**, 9357–63. 98

KINNAMAN, T.C. (2011). The economic impact of shale gas extraction: A review of existing studies. *Ecological Economics*, **70**, 1243–1249. 100

KLOSS, C., NEWLAND, M., ORAM, D., FRASER, P., BRENNINKMEIJER, C., RÖCKMANN, T. & LAUBE, J. (2014). Atmospheric Abundances, Trends and Emissions of CFC-216ba, CFC-216ca and HCFC-225ca. *Atmosphere*, **5**, 420–434. 8

KNIGHTON, W., FORTNER, E., MIDEY, A.J., VIGGIANO, A., HERNDON, S., WOOD, E. & KOLB, C. (2009). HCN detection with a proton transfer reaction mass spectrometer. *International Journal of Mass Spectrometry*, **283**, 112–121. 69

KOSUSKO, M. & NUNEZ, C. (1990). Destruction of Volatile Organic Compounds Using Catalytic Oxidation. *Journal of the Air & Waste Management Association*, **40**, 254–259. 46

KRÜGER, O., KALBE, U., MEISSNER, K. & SOBOTTKA, S. (2014). Sorption effects interfering with the analysis of polycyclic aromatic hydrocarbons (PAH) in aqueous samples. *Talanta*, **122**, 151–6. 44

LAI, S.C., BAKER, A.K., SCHUCK, T.J., VAN VELTHOVEN, P., ORAM, D.E., ZAHN, A., HERMANN, M., WEIGELT, A., SLEMR, F., BRENNINKMEIJER, C.A.M. & ZIEREIS, H. (2010). Pollution events observed during CARIBIC flights in the upper troposphere between South China and the Philippines. *Atmospheric Chemistry and Physics*, **10**, 1649–1660. 84

LAUBE, J.C., NEWLAND, M.J., HOGAN, C., BRENNINKMEIJER, C.A.M., FRASER, P.J., MARTINERIE, P., ORAM, D.E., REEVES, C.E., RÖCKMANN, T., SCHWANDER, J., WITRANT, E. & STURGES, W.T. (2014). Newly detected ozone-depleting substances in the atmosphere. *Nature Geoscience*, **7**, 266–269. 8

LELIEVELD, J., CRUTZEN, P.J., RAMANATHAN, V., ANDREAE, M.O., BRENNINKMEIJER, C.M., CAMPOS, T., CASS, G.R., DICKERSON, R.R., FISCHER, H., DE GOUW, J.A., HANSEL, A., JEFFERSON, A., KLEY, D., DE LAAT, A.T., LAL, S., LAWRENCE, M.G., LOBERT, J.M., MAYOL-BRACERO, O.L., MITRA, A.P., NOVAKOV, T., OLTMANS, S.J., PRATHER, K.A., REINER, T., RODHE, H., SCHEEREN, H.A., SIKKA, D. & WILLIAMS, J. (2001). The Indian Ocean experiment: widespread air pollution from South and Southeast Asia. *Science (New York, N.Y.)*, **291**, 1031–1036. 88

LELIEVELD, J., BUTLER, T.M., CROWLEY, J.N., DILLON, T.J., FISCHER, H., GANZEVELD, L., HARDER, H., LAWRENCE, M.G., MARTINEZ, M., TARABORRELLI, D. & WILLIAMS, J. (2008). Atmospheric oxidation capacity sustained by a tropical forest. *Nature*, **452**, 737–740. 8, 9

LEWIS, A.C., HOPKINS, J.R., CARPENTER, L.J., STANTON, J., READ, K.A. & PILLING, M.J. (2005). Sources and sinks of acetone, methanol, and acetaldehyde in North Atlantic marine air. *Atmospheric Chemistry and Physics*, **5**, 1963–1974. 5

LI, R., PALM, B.B., BORBON, A., GRAUS, M., WARNEKE, C., ORTEGA, A.M., DAY, D.A., BRUNE, W.H., JIMENEZ, J.L. & DE GOUW, J.A. (2013). Laboratory studies on secondary organic aerosol formation from crude oil vapors. *Environmental Science & Technology*, **47**, 12566–74. 10, 146

LI, R., WARNEKE, C., GRAUS, M., FIELD, R., GEIGER, F., VERES, P.R., SOLTIS, J., LI, S.M., MURPHY, S.M., SWEENEY, C., PÉTRON, G., ROBERTS, J.M. & DE GOUW, J. (2014). Measurements of hydrogen sulfide (H2S) using PTR-MS: calibration, humidity dependence, inter-comparison and results from field studies in an oil and gas production region. *Atmospheric Measurement Techniques Discussions [accepted for final publication in AMT: 18.9.2014]*, **7**, 6205–6243. 19, 69, 70, 71, 149, 177, 178

LINDINGER, W. & JORDAN, A. (1998). Proton-transfer-reaction mass spectrometry (PTRMS): on-line monitoring of volatile organic compounds at pptv levels. *Chemical Society Reviews*, **27**, 347. 11

LINDINGER, W., HIRBER, J. & PARETZKE, H. (1993). An ion/molecule-reaction mass spectrometer used for on-line trace gas analysis. *International Journal of Mass Spectrometry and Ion Processes*, **129**, 79–88. 11

LINDINGER, W., HANSEL, A. & JORDAN, A. (1998). On-line monitoring of volatile organic compounds at pptv levels by means of proton-transfer-reaction mass spectrometry (PTR-MS) medical applications, food control and environmental research. *International Journal of Mass Spectrometry and Ion Processes*, **173**, 191–241. 11, 15, 89

LINSTROM, P.J. & MALLARD, W.G. (2001). The NIST Chemistry WebBook: A Chemical Data Resource on the Internet. *Journal of Chemical & Engineering Data*, **46**, 1059–1063. 17

LINSTROM, P.J. & MALLARD, W.G., eds. (2011). *NIST Chemistry WebBook, NIST Standard Reference Database Number 69*. United States National Institute of Standards and Technology (NIST). 17, 19

LITOVITZ, A., CURTRIGHT, A., ABRAMZON, S., BURGER, N. & SAMARAS, C. (2013). Estimation of regional air-quality damages from Marcellus Shale natural gas extraction in Pennsylvania. *Environmental Research Letters*, **8**, 014017. 98

LIU, D., FEILBERG, A., ADAMSEN, A.P. & JONASSEN, K.E. (2011). The effect of slurry treatment including ozonation on odorant reduction measured by in-situ PTR-MS. *Atmospheric Environment*, **45**, 3786–3793. 149

MANUEL, J. (2010). EPA tackles fracking. *Environmental Health Perspectives*, **118**, A199. 101

MARCH, R. (1997). An introduction to quadrupole ion trap mass spectrometry. *Journal of Mass Spectrometry*, **32**, 351–369. 12

MARTIN, R., MOORE, K., MANSFIELD, M., HILL, S., HARPER, K. & SHORTHILL, H. (2011). Final Report: Uinta Basin Winter Ozone and Air Quality Study December 2010 - March 2011. Tech. rep., Energy Dynamics Laboratory, Utah State University Research Foundation (USURF). 100, 106

MATSUMI, Y. & KAWASAKI, M. (2003). Photolysis of atmospheric ozone in the ultraviolet region. *Chemical Reviews*, **103**, 4767–82. 6

MCFARLAND, M. (1973a). Flow-drift technique for ion mobility and ion-molecule reaction rate constant measurements. I. Apparatus and mobility measurements. *The Journal of Chemical Physics*, **59**, 6610. 11

MCFARLAND, M. (1973b). Flow-drift technique for ion mobility and ion-molecule reaction rate constant measurements. II. Positive ion reactions of N+, O+, and H2+ with O2 and O+ with N2 from thermal to ~2 eV. *The Journal of Chemical Physics*, **59**, 6620–6628. 11

MCFARLAND, M. (1973c). Flow-drift technique for ion mobility and ion-molecule reaction rate constant measurements. III. Negative ion reactions of O with CO, NO, H2, and D2. *The Journal of Chemical Physics*, **59**, 6629. 11

MCILVAINE, R. & JAMES, A. (2010). The potential of gas shale. *World Pumps*, **2010**, 16–18. 100

MCKENZIE, L.M., WITTER, R.Z., NEWMAN, L.S. & ADGATE, J.L. (2012). Human health risk assessment of air emissions from development of unconventional natural gas resources. *Science of the Total Environment*, **424**, 79–87. 98, 100

METTS, T.A. & BATTERMAN, S.A. (2006). Effect of VOC loading on the ozone removal efficiency of activated carbon filters. *Chemosphere*, **62**, 34–44. 45

MIDDLEBROOK, A.M., MURPHY, D.M., AHMADOV, R., ATLAS, E.L., BAHREINI, R., BLAKE, D.R., BRIOUDE, J., DE GOUW, J.A., FEHSENFELD, F.C., FROST, G.J., HOLLOWAY, J.S., LACK, D.A., LANGRIDGE, J.M., LUEB, R.A., MCKEEN, S.A., MEAGHER, J.F., MEINARDI, S., NEUMAN, J.A., NOWAK, J.B., PARRISH, D.D., PEISCHL, J., PERRING, A.E., POLLACK, I.B., ROBERTS, J.M., RYERSON, T.B., SCHWARZ, J.P., SPACKMAN, J.R., WARNEKE, C. & RAVISHANKARA, A.R. (2012). Air quality implications of the Deepwater Horizon oil spill. *Proceedings of the National Academy of Sciences of the United States of America.* 10, 146

MILLER, P.E. & DENTON, M.B. (1986). The quadrupole mass filter: Basic operating concepts. *Journal of Chemical Education*, **63**, 617. 29, 30

MOHAMED, M.F., KANG, D. & ANEJA, V.P. (2002). Volatile organic compounds in some urban locations in United States. *Chemosphere*, **47**, 863–882. 116

MOHR, P.J., TAYLOR, B.N. & NEWELL, D.B. (2012). CODATA recommended values of the fundamental physical constants: 2010. *Reviews of Modern Physics*, **84**, 1527–1605. 21

MONKS, P.S. (2005). Gas-phase radical chemistry in the troposphere. *Chemical Society Reviews*, **34**, 376–95. 9

MOORE, C.W., ZIELINSKA, B., PÉTRON, G. & JACKSON, R.B. (2014). Air impacts of increased natural gas acquisition, processing, and use: a critical review. *Environmental Science & Technology*, **48**, 8349–59. 100

MPIC (2012). V25-Bedienungshandbuch. Tech. rep., Max Planck Institute for Chemistry (MPIC), Electronics Group. 40, 43

MÜLLER, M., GRAUS, M., RUUSKANEN, T.M., SCHNITZHOFER, R., BAMBERGER, I., KASER, L., TITZMANN, T., HÖRTNAGL, L., WOHLFAHRT, G., KARL, T. & HANSEL, A. (2010). First eddy covariance flux measurements by PTR-TOF. *Atmospheric Measurement Techniques*, **3**, 387–395. 72

MUNSON, M.S.B. & FIELD, F.H. (1966). Chemical Ionization Mass Spectrometry. I. General Introduction. *Journal of the American Chemical Society*, **88**, 2621–2630. 11

NEUMAIER, M., RUHNKE, R., KIRNER, O., ZIEREIS, H., STRATMANN, G., BRENNINKMEIJER, C.A.M. & ZAHN, A. (2014). Impact of acetone (photo)oxidation on HOx production in the UT/LMS based on CARIBIC passenger aircraft observations and EMAC simulations. *Geophysical Research Letters*, **41**, 3289–3297. 9, 84, 89

OLVER, F.W.J., LOZIER, D.W., BOISVERT, R.F. & CLARK, C.W. (2010). *NIST Handbook of Mathematical Functions*. Cambridge University Press. 29

ORLANDO, J.J. & TYNDALL, G.S. (2012). Laboratory studies of organic peroxy radical chemistry: an overview with emphasis on recent issues of atmospheric significance. *Chemical Society Reviews*, **41**, 6294–6317. 7

OSBORN, S.G., VENGOSH, A., WARNER, N.R. & JACKSON, R.B. (2011). Methane contamination of drinking water accompanying gas-well drilling and hydraulic fracturing. *Proceedings of the National Academy of Sciences of the United States of America*, **108**, 8172–6. 100

O'SULLIVAN, F. & PALTSEV, S. (2012). Shale gas production: potential versus actual greenhouse gas emissions. *Environmental Research Letters*, **7**, 044030. 100

PÉTRON, G., FROST, G., MILLER, B.R., HIRSCH, A.I., MONTZKA, S.A., KARION, A., TRAINER, M., SWEENEY, C., ANDREWS, A.E., MILLER, L., KOFLER, J., BAR-ILAN, A., DLUGOKENCKY, E.J., PATRICK, L., MOORE, C.T., RYERSON, T.B., SISO, C., KOLODZEY, W., LANG, P.M., CONWAY, T., NOVELLI, P., MASARIE, K., HALL, B., GUENTHER, D., KITZIS, D., MILLER, J., WELSH, D., WOLFE, D., NEFF, W. & TANS, P. (2012). Hydrocarbon emissions characterization in the Colorado Front Range: A pilot study. *Journal of Geophysical Research*, **117**, 1–19. 98

PFEIFFER-VACUUM (2007). Operating Instructions QMG 422 BG 805 981 BE. Tech. Rep. 0112, Pfeiffer Vacuum GmbH, D-35614 Asslar. 51

PINTO, J. (2009). Atmospheric Chemistry: Wyoming winter smog. *Nature Geoscience*, **2**, 88–89. 8, 100

PÖSCHL, U. (2005). Atmospheric aerosols: composition, transformation, climate and health effects. *Angewandte Chemie (International ed. in English)*, **44**, 7520–40. 10

PRAZELLER, P., PALMER, P.T., BOSCAINI, E., JOBSON, T. & ALEXANDER, M. (2003). Proton transfer reaction ion trap mass spectrometer. *Rapid Communications in Mass Spectrometry*, **17**, 1593–9. 12

PRINN, R.G. (2003). THE CLEANSING CAPACITY OF THE ATMOSPHERE. *Annual Review of Environment and Resources*, **28**, 29–57. 8

RAHM, D. (2011). Regulating hydraulic fracturing in shale gas plays: The case of Texas. *Energy Policy*, **39**, 2974–2981. 100

RANDEL, W.J. & JENSEN, E.J. (2013). Physical processes in the tropical tropopause layer and their roles in a changing climate. *Nature Geoscience*, **6**, 169–176. 84

RAPPENGLÜCK, B., ACKERMANN, L., ALVAREZ, S., GOLOVKO, J., BUHR, M., FIELD, R.A., SOLTIS, J., MONTAGUE, D.C., HAUZE, B., ADAMSON, S., RISCH, D., WILKERSON, G., BUSH, D., STOECKENIUS, T. & KESLAR, C. (2014). Strong wintertime ozone events in the Upper Green River basin, Wyoming. *Atmospheric Chemistry and Physics*, **14**, 4909–4934. 8

RICE, D.D., FOUCH, T.D. & JOHNSON, R.C. (1992). Influence of source rock type, thermal maturity, and migration on composition and distribution of natural gases, Uinta Basin, Utah. *Hydrocarbon and Mineral Resources of the Uinta basin, Utah and Colorado: Utah Geological Assocation Guidebook 20*, 95–110. 99

RIIPINEN, I., YLI-JUUTI, T., PIERCE, J.R., PETÄJÄ, T., WORSNOP, D.R., KULMALA, M. & DONAHUE, N.M. (2012). The contribution of organics to atmospheric nanoparticle growth. *Nature Geoscience*, **5**, 453–458. 10

ROBINSON, A.L., DONAHUE, N.M., SHRIVASTAVA, M.K., WEITKAMP, E.A., SAGE, A.M., GRIESHOP, A.P., LANE, T.E., PIERCE, J.R. & PANDIS, S.N. (2007). Rethinking organic aerosols: semivolatile emissions and photochemical aging. *Science (New York, N.Y.)*, **315**, 1259–62. 9, 10

RYERSON, T.B., AIKIN, K.C., ANGEVINE, W.M., ATLAS, E.L., BLAKE, D.R., BROCK, C.A., FEHSENFELD, F.C., GAO, R.S., DE GOUW, J.A., FAHEY, D.W., HOLLOWAY, J.S., LACK, D.A., LUEB, R.A., MEINARDI, S., MIDDLEBROOK, A.M., MURPHY, D.M., NEUMAN, J.A., NOWAK, J.B., PARRISH, D.D., PEISCHL, J., PERRING, A.E., POLLACK, I.B., RAVISHANKARA, A.R., ROBERTS, J.M., SCHWARZ, J.P., SPACKMAN, J.R., STARK, H., WARNEKE, C. & WATTS, L.A. (2011). Atmospheric emissions from the Deepwater Horizon spill constrain air-water partitioning, hydrocarbon fate, and leak rate. *Geophysical Research Letters*, **38**, L07803. 146

RYERSON, T.B., CAMILLI, R., KESSLER, J.D., KUJAWINSKI, E.B., REDDY, C.M., VALENTINE, D.L., ATLAS, E., BLAKE, D.R., DE GOUW, J., MEINARDI, S., PARRISH, D.D., PEISCHL, J., SEEWALD, J.S. & WARNEKE, C. (2012). Chemical data quantify Deepwater Horizon hydrocarbon flow rate and environmental distribution. *Proceedings of the National Academy of Sciences of the United States of America*, 1–8. 146

RYERSON, T.B., ANDREWS, A.E., ANGEVINE, W.M., BATES, T.S., BROCK, C.A., CAIRNS, B., COHEN, R.C., COOPER, O.R., DE GOUW, J.A., FEHSENFELD, F.C., FERRARE, R.A., FISCHER, M.L., FLAGAN, R.C., GOLDSTEIN, A.H., HAIR, J.W., HARDESTY, R.M., HOSTETLER, C.A., JIMENEZ, J.L., LANGFORD, A.O., MCCAULEY, E., MCKEEN, S.A., MOLINA, L.T., NENES, A., OLTMANS, S.J., PARRISH, D.D., PEDERSON, J.R., PIERCE, R.B., PRATHER, K., QUINN, P.K., SEINFELD, J.H., SENFF, C.J., SOROOSHIAN, A., STUTZ, J., SURRATT, J.D., TRAINER, M., VOLKAMER, R., WILLIAMS, E.J. & WOFSY, S.C. (2013). The 2010 California Research at the Nexus of Air Quality and Climate Change (CalNex) field study. *Journal of Geophysical Research: Atmospheres*, **118**, 5830–5866. 10

SCHEELE, M.P., SIEGMUND, P.C. & VAN VELTHOVEN, P.F.J. (1996). Sensitivity of trajectories to data resolution and its dependence on the starting point: in or outside a tropopause fold. *Meteorological Applications*, **3**, 267–273. 84

SCHEEREN, H.A., LELIEVELD, J., ROELOFS, G.J., WILLIAMS, J., FISCHER, H., DE REUS, M., DE GOUW, J.A., WARNEKE, C., HOLZINGER, R., SCHLAGER, H., KLÜPFEL, T., BOLDER, M., VAN DER VEEN, C. & LAWRENCE, M. (2003). The impact of monsoon outflow from India and Southeast Asia in the upper troposphere over the eastern Mediterranean. *Atmospheric Chemistry and Physics*, **3**, 1589–1608. 88

SCHNEIDER, M. & HASE, F. (2011). Optimal estimation of tropospheric H2O and δD with IASI/METOP. *Atmospheric Chemistry and Physics*, **11**, 11207–11220. 169

SCHNELL, R.C., OLTMANS, S.J., NEELY, R.R., ENDRES, M.S., MOLENAR, J.V. & WHITE, A.B. (2009). Rapid photochemical production of ozone at high concentrations in a rural site during winter. *Nature Geoscience*, **2**, 120–122. 8, 99

SCHUMANN, U. & ANDREAE, M.O. (2006). HALO eröffnet neue Perspektiven in der Atmosphärenforschung: Höher, grösser, weiter. *DLR Nachrichten*, 37–39. 86

SEINFELD, J.J.H. & PANDIS, S.N.S. (2006). *Atmospheric Chemistry and Physics: From Air Pollution to Climate Change*. 3, 4, 18

SINGH, H.B., KANAKIDOU, M., CRUTZEN, P.J. & JACOB, D.J. (1995). High concentrations and photochemical fate of oxygenated hydrocarbons in the global troposphere. *Nature*, **378**, 50–54. 8

SKRTIC, L. (2006). *Hydrogen Sulfide, Oil and Gas, and People's Health*. Ph.D. thesis, University of California, Berkeley. 151

SLEVIN, P.J. & HARRISON, W.W. (1975). The Hollow Cathode Discharge as a Spectrochemical Emission Source. *Applied Spectroscopy Reviews*, **10**, 201–255. 16, 47

SPRUNG, D. (2002). *Aufbau und Anwendung eines mittels chemischer Ionisation unter Atmosphärendruck arbeitenden Quadrupol Massenspektrometers (AP-CIMS) für schnelle Spurengasmessungen*. Ph.D. thesis, Brandenburgische Technische Universität Cottbus. 28

SPRUNG, D. & ZAHN, A. (2010). Acetone in the upper troposphere/lowermost stratosphere measured by the CARIBIC passenger aircraft: Distribution, seasonal cycle, and variability. *Journal of Geophysical Research*, **115**, D16301. 35, 84

SPRUNG, D., JOST, C., REINER, T., HANSEL, A. & WISTHALER, A. (2001). Acetone and acetonitrile in the tropical Indian Ocean boundary layer and free troposphere: Aircraft-based intercomparison of AP-CIMS and PTR-MS measurements. *Journal of Geophysical Research*, **106**, 28511. 84

SULZER, P., EDTBAUER, A., HARTUNGEN, E., JÜRSCHIK, S., JORDAN, A., HANEL, G., FEIL, S., JAKSCH, S., MÄRK, L. & MÄRK, T.D. (2012). From conventional proton-transfer-reaction mass spectrometry (PTR-MS) to universal trace gas analysis. *International Journal of Mass Spectrometry*, **321-322**, 66–70. 12

SULZER, P., HARTUNGEN, E., HANEL, G., FEIL, S., WINKLER, K., MUTSCHLECHNER, P., HAIDACHER, S., SCHOTTKOWSKY, R., GUNSCH, D., SEEHAUSER, H., STRIEDNIG, M., JÜRSCHIK, S., BREIEV, K., LANZA, M., HERBIG, J., MÄRK, L., MÄRK, T.D. & JORDAN, A. (2014). A Proton Transfer Reaction-Quadrupole interface Time-Of-Flight Mass Spectrometer (PTR-QiTOF): High speed due to extreme sensitivity. *International Journal of Mass Spectrometry*, **368**, 1–5. 12

TANAKA, K., MACKAY, G.I. & BOHME, D.K. (1978). Rate and equilibrium constant measurements for gas-phase proton-transfer reactions involving H2O, H2S, HCN, and H2CO. *Canadian Journal of Chemistry*, **56**, 193–204. 69

TARVER, G.A. & DASGUPTA, P.K. (1997). Oil Field Hydrogen Sulfide in Texas: Emission Estimates and Fate. *Environmental Science & Technology*, **31**, 3669–3676. 149

UNGER, N. (2014). Human land-use-driven reduction of forest volatiles cools global climate. *Nature Climate Change*, **4**, 907–910.

VERES, P., GILMAN, J.B., ROBERTS, J.M., KUSTER, W.C., WARNEKE, C., BURLING, I.R. & DE GOUW, J. (2010). Development and validation of a portable gas phase standard generation and calibration system for volatile organic compounds. *Atmospheric Measurement Techniques*, **3**, 683–691. 58

VIEHLAND, L. & MASON, E. (1995). Transport Properties of Gaseous Ions over a Wide Energy Range, IV. *Atomic Data and Nuclear Data Tables*, **60**, 37–95. 20

VIGGIANO, A.A. & MORRIS, R.A. (1996). Rotational and Vibrational Energy Effects on IonMolecule Reactivity As Studied by the VT-SIFDT Technique. *The Journal of Physical Chemistry*, **100**, 19227–19240. 20, 21

VLASENKO, A., MACDONALD, A.M., SJOSTEDT, S.J. & ABBATT, J.P.D. (2010). Formaldehyde measurements by Proton transfer reaction Mass Spectrometry (PTR-MS): correction for humidity effects. *Atmospheric Measurement Techniques*, **3**, 1055–1062. 19, 65

WARNEKE, C. (2004). Comparison of daytime and nighttime oxidation of biogenic and anthropogenic VOCs along the New England coast in summer during New England Air Quality Study 2002. *Journal of Geophysical Research*, **109**, D10309. 5

WARNEKE, C., VAN DER VEEN, C., LUXEMBOURG, S., DE GOUW, J. & KOK, A. (2001). Measurements of benzene and toluene in ambient air using proton-transfer-reaction mass spectrometry: calibration, humidity dependence, and field intercomparison. *International Journal of Mass Spectrometry*, **207**, 167–182. 23, 65

WARNEKE, C., DE GOUW, J.A., KUSTER, W.C., GOLDAN, P.D. & FALL, R. (2003). Validation of Atmospheric VOC Measurements by Proton-Transfer- Reaction Mass Spectrometry Using a Gas-Chromatographic Preseparation Method. *Environmental Science & Technology*, **37**, 2494–2501. 23, 58

WARNEKE, C., GOUW, J.A., LOVEJOY, E.R., MURPHY, P.C., KUSTER, W.C. & FALL, R. (2005). Development of proton-transfer ion trap-mass spectrometry: on-line detection and identification of volatile organic compounds in air. *Journal of the American Society for Mass Spectrometry*, **16**, 1316–1324. 12, 73

WARNEKE, C., MCKEEN, S.A., DE GOUW, J.A., GOLDAN, P.D., KUSTER, W.C., HOLLOWAY, J.S., WILLIAMS, E.J., LERNER, B.M., PARRISH, D.D., TRAINER, M., FEHSENFELD, F.C., KATO, S., ATLAS, E.L., BAKER, A. & BLAKE, D.R. (2007). Determination of urban volatile organic compound emission ratios and comparison with an emissions database. *Journal of Geophysical Research*, **112**, D10S47. 4

WARNEKE, C., ROBERTS, J., VERES, P., GILMAN, J., KUSTER, W., BURLING, I., YOKELSON, R. & DE GOUW, J. (2011a). VOC identification and inter-comparison from laboratory biomass burning using PTR-MS and PIT-MS. *International Journal of Mass Spectrometry*, **303**, 6–14. 84

WARNEKE, C., VERES, P., HOLLOWAY, J.S., STUTZ, J., TSAI, C., ALVAREZ, S., RAPPENGLUECK, B., FEHSENFELD, F.C., GRAUS, M., GILMAN, J.B. & DE GOUW, J.A. (2011b). Airborne formaldehyde measurements using PTR-MS: calibration, humidity dependence, inter-comparison and initial results. *Atmospheric Measurement Techniques*, **4**, 2345–2358. 19, 65, 68

WARNEKE, C., DE GOUW, J.A., EDWARDS, P.M., HOLLOWAY, J.S., GILMAN, J.B., KUSTER, W.C., GRAUS, M., ATLAS, E., BLAKE, D., GENTNER, D.R., GOLDSTEIN, A.H., HARLEY, R.A., ALVAREZ, S., RAPPENGLUECK, B., TRAINER, M. & PARRISH, D.D. (2013). Photochemical aging of volatile organic compounds in the Los Angeles basin: Weekday-weekend effect. *Journal of Geophysical Research: Atmospheres*, **118**, 5018–5028. 116

WARNEKE, C., GEIGER, F., EDWARDS, P.M., DUBE, W., PÉTRON, G., KOFLER, J., ZAHN, A., BROWN, S.S., GRAUS, M., GILMAN, J., LERNER, B., PEISCHL, J., RYERSON, T.B., DE GOUW, J.A. & ROBERTS, J.M. (2014). Volatile organic compound emissions from the oil and natural gas industry in the Uinta Basin, Utah: point sources compared to ambient air composition. *Atmospheric Chemistry and Physics Discussions [accepted for final publication in ACP: 6.9.2014]*, **14**, 11895–11927. 64, 75, 99, 117, 118, 119, 121, 127, 128, 130, 131

WILLIAMS, J. (2004). Organic Trace Gases in the Atmosphere: An Overview. *Environmental Chemistry*, **1**, 125. 4

WISTHALER, A., APEL, E.C., BOSSMEYER, J., HANSEL, A., JUNKERMANN, W., KOPPMANN, R., MEIER, R., MÜLLER, K., SOLOMON, S.J., STEINBRECHER, R., TILLMANN, R. & BRAUERS, T. (2008). Technical Note: Intercomparison of formaldehyde measurements at the atmosphere simulation chamber SAPHIR. *Atmospheric Chemistry and Physics*, **8**, 2189–2200. 68

YUAN, B., WARNEKE, C., SHAO, M. & DE GOUW, J.A. (2014). Interpretation of volatile organic compound measurements by proton-transfer-reaction mass spectrometry over the deepwater horizon oil spill. *International Journal of Mass Spectrometry*, **358**, 43–48. 58, 64, 116, 122, 123, 146

ZAHN, A. & BRENNINKMEIJER, C.A. (2003). New Directions: A Chemical Tropopause Defined. *Atmospheric Environment*, **37**, 439–440. 83

ZAHN, A., WEPPNER, J., WIDMANN, H., SCHLOTE-HOLUBEK, K., BURGER, B., KÜHNER, T. & FRANKE, H. (2012). A fast and precise chemiluminescence ozone detector for eddy flux and airborne application. *Atmospheric Measurement Techniques*, **5**, 363–375. 40, 89

ZHANG, Y., GABLE, C.W., ZYVOLOSKI, G.A. & WALTER, L.M. (2009). Hydrogeochemistry and gas compositions of the Uinta Basin: A regional-scale overview. *AAPG Bulletin*, **93**, 1087–1118. 99

ZIEMANN, P.J. & ATKINSON, R. (2012). Kinetics, products, and mechanisms of secondary organic aerosol formation. *Chemical Society Reviews*, **41**, 6582–6605. 10

Nomenclature

AMP	Amplifier
API	American Petroleum Institute
B	Gas-phase Basicity
CAP	Capillary
CARIBIC	Civil Aircraft for the Regular Investigation of the atmosphere Based on an Instrument Container
CAT	Catalyst
CCF	Charcoal Filter
CFC	Chlorofluorocarbon
CI(MS)	Chemical Ionization (Mass Spectrometry)
cps	counts per second
CRDS	Cavity Ring Down Spectroscopy
CSD	Chemical Science Division
CTR	Charge-Transfer-Reaction
DT	Drift tube
E	Electrical Field
e	Elementary charge: $1.60217733(49) \times 10^{-19}\,C$
ECMWF	European Centre for Medium-Range Weather Forecasts
EMI	ElectroMagnetic Interference
ER	Enhancement Ratio
ESRL	Earth System Research Laboratory
ex-FT	Extra-tropical free troposphere
GC-MS	Gas Chromatography Mass Spectrometer
GMD	Global Monitoring Division
h	Planck constant: $h = 6.62606876(52) \times 10^{-34}\,J\cdot s$
HALO	High Altitude and LOng Range
I	Ion Count Rate
ICAO	International Civil Aviation Organization
ID	Inner Diameter
IPCC	Intergovernmental Panel on Climate Change

IS	Ion Source
ISDR	Ion Source Drift Region
IVC	Intermediate Vacuum Chambers
k	rate coefficient
LOD	Limit of Detection
m	Mass
MFC	Mass Flow Controller
MPIC	Max Planck Institute for Chemistry
MUSICA	MUlti-platform remote Sensing of Isotopologues for investigating the Cycle of Atmospheric water
N	Gas Number Density
NAAQS	National Ambient Air Quality Standards
NC	Nose Cone
ncps	normalized counts per second
NIST	National Institute of Standards and Technology
NMVOC	Non-methane Volatile Organic Compounds
NOAA	National Oceanic And Atmospheric Administration
O.x	Orifice
O&NG	Oil and Natural Gas
OD	Outer Diameter
OMO	Oxidation Mechanism Observations
P	Pressure
P.x	Pressure sensor
PA	Proton Affinity
PBL	Planetary Boundary Layer
PC	Pressure Controller
PEEK	PolyEther Ether Ketone
PFA	Perfluoroalkoxy
PM	Particulate Mass
ppb(V)	parts per billion (by volume) (10^{-9})
ppm(V)	parts per million (by volume) (10^{-6})
ppq(V)	parts per quadrillion (by volume) (10^{-15})
ppt(V)	parts per trillion (by volume) (10^{-12})
PTR(-MS)	Proton-Transfer-Reaction (Mass Spectrometry/Spectrometer)
PTR-IF-MS	Proton-Transfer-Reaction Ion Funnel Mass Spectrometer
q	Ion Charge

QMS	Quadrupole Mass Spectrometry/Spectrometer
R	organic hydrocarbon species
r_0	Radius
RF	Radio Frequency
RPDB	Rack Power Distribution Box
S	Sensitivity/Calibration factor
S/N	Signal-to-noise ratio
SEM	Secondary Electron Multiplier
SOA	Secondary Organic Aerosol
T	Temperature; Transmission Factors
t	Time
TGI	Trace Gas Inlet (HALO)
TMP	Turbo Molecular Pump
ToF	Time of Flight
ULW-PTR-MS	Ultra-Light-Weight Proton-Transfer-Reaction Mass Spectrometer
US EPA	United States Environmental Protection Agency
UT/LMS	Upper Troposphere and Lowermost Stratosphere
UV	Ultraviolet
V.x	Valve
VMR	Volume Mixing Ratio
VOC	Volatile Organic Compound
WRAP	Western Regional Air Partnership
z	Charge/number of elementary charges
ϵ	Sensitivity
λ	Mean free path
μ	Ion Mobility
ν	Frequency
ω	Frequency
ϕ	Potential
σ	Standard Deviation
τ	Time

Acknowledgments

This work was carried out between October 2010 and October 2014 in the group for In Situ Measurements Onboard Aircraft at the Institute for Meteorology and Climate Research - Atmospheric Trace Gases and Remote Sensing (IMK-ASF); Department of Physics; Karlsruhe Institute of Technology (KIT) in Karlsruhe/Germany. Financial support was provided (within others) by the German Research Foundation (DFG) and IAGOS-D (pilot-phase). The research visit at the Earth System Research Laboratory (ESRL); National Oceanic and Atmospheric Administration (NOAA) in Boulder, Colorado/USA, was funded by a scholarship of the Graduate School for Climate and Environment (GRACE); Helmholtz Association (HGF).

Here I want to acknowledge the people that commendably gave their kind assistance and support during the preparation of the presented doctoral thesis:

First and foremost, I wish to sincerely thank my doctoral supervisor and institute director **Prof. Dr. Johannes Orphal** (KIT IMK-ASF) for his motivating and personal guidance and steady interest during my work on this exciting topic.

I am very thankful to my second reviewer **Prof. Dr. Thomas Leisner** (KIT IMK-AAF) for his kind participation in my doctoral committee and very welcome independent advices in our personal conversations.

With much appreciation I want to express my gratitude to current and former group members and colleagues from other institutions for all discussions about my research topic and their valuable suggestions on my work, any technical assistances in the lab and during the field/aircraft missions, their help with programming and data processing, administrative support, personal guidance and for a great atmosphere in- and outside the lab:
Dr. Joel Ferreira de Brito, Derya Cayiroglu, Dr. Emanuel Christner, Dr.-Ing. Christoph Dyroff, Dr. Felix Friederich, Dr. Ottmar Möhler, Dr. Marco Neumaier, Antoine Petrelli, Benjamin Riexinger, Alexander Streili, Helmut Widmann, Astrud Wiessner, Dr. Wolfgang Woiwode and **Dr. Andreas Zahn** (all from IMK),
Dr.-Ing. Christian Lucas (from GRACE),
Michael Flanz, Christian Gurk and **Dr. Frank Helleis** (all from Max Planck Institute for Chemistry, Mainz/Germany),

Boris Schneider (from Enviscope, Frankfurt/Germany),
Prof. Dr. Richard Grant (from Purdue University, West Lafayette, Indiana/USA),
Kenneth Aikin, Jane August, Dr. Steven Brown, Dr. Joost de Gouw, William Dubé, Dr. Peter Edwards, Dr. Jessica Gilman, Dr. Martin Graus, Jonathan Kofler, Dr. Brian Lerner, Dr. Gabrielle Pétron, Dr. James M. Roberts and **Dr. Carsten Warneke** (all from NOAA).

I gratefully acknowledge my supervisor and group leader **Dr. Andreas Zahn** for the opportunities, the large scope for development and the independency I was allowed to experience during my thesis. I appreciate his trust in my work that made room for self-development and learning new things.

Many thanks to **Dr. Frank Helleis** for a great collaboration during the development phase of the ULW-PTR-MS. Without him, his professional guidance and personal engagement the presented system would not exist. I highly appreciate the uncounted jointly spent hours at the lab in Mainz or elsewhere in both professional and personal respect.

Many thanks to **Dr. Carsten Warneke** for the unique opportunity of being visiting scientist at his lab at NOAA and the fantastic chance to take an active part in his team during the presented field measurements. I very much appreciate his scientific and personal support and greatly enjoyed the close and trusting cooperation. My time at NOAA was a great pleasure.

Last but not least, I want to thank my family and my friends for their encouragements and emotional support during the years of working on this thesis. My personal very special thanks go to my wife **Katharina Geiger** for her understanding, all good advices and her extraordinary and enduring support.